Take the Next Step in Your IT Career

CompTIA®
Project+® Practice Tests
Exam PK0-005
Second Edition

CompTIA®
Project+® Practice Tests
Exam PK0-005
Second Edition

Brett J. Feddersen

SYBEX®
A Wiley Brand

This book is dedicated to my family, who without them, I would never know the Wi-Fi was not working, someone needs to go to the grocery store, when allowance day is, and most importantly, that hockey is not life and watching the Colorado Avalanche lift the Stanley Cup for the third time is not a crying moment.

I would also not know what it is like to have my home light up my life, the joy of laughter brought on by the antics of stuffed animals, watching your greatest creations grow smarter, stronger, and more independent. Alas, I would not know the most driven, intelligent, and caring woman in the world dedicated to making every day better for those around her. I look forward to calling you Dr. Pratt soon.

Acknowledgments

Holy cow! Do you know what is involved in writing a book? The team of dedicated professionals who helped to create this resource for you in order to advance your project management career is the best in the business. It is due to the efforts of the following individuals that the quality of this book is strong.

The spark for this effort began with Kenyon Brown, the senior acquisition editor who assembled this great team. Liz Britten, the project manager, helped guide and drive the completion of the book through COVID-19 times. Her efforts made this tool the very best product it could be to assist you in passing the CompTIA Project+ exam. Vanina Mangano helped to ensure the technical viability of this book. Her talents and dedication in the project management field are globally known, and she is a terrific collaboration partner. Liz Welch did an amazing job as the copy editor in helping to make sure the writing made sense, was consistent, and grammatically correct. The rest of the production team, led by Christine O'Connor, made sure that this text made it into print in a readable form. Thank you all for your collaboration efforts!

I would also like to acknowledge Kim Heldman, who has been a professional mentor to me for close to two decades. It was Kim, one of the most courageous leaders I have ever met, who got me started down the path of project management certification, authorship, and speaking. Kim is one of the great leaders in public sector information technology. She brings order and structure to a scary and complex area.

No acknowledgment has any validity without honoring one's parents. They showed me *Animal House* when I was nine years old as a documentary of their days in college, and they helped to create a worldview in which you do not take yourself so damn seriously, be kind to your neighbor, to laugh freely, and to love with your whole being. I love you, Pete and Madeline.

Most importantly, I would like to thank my wife, April, and my children, Kayla, Marcus, and Adric. They endured the opportunity costs of not building LEGOs, helping with homework, or watching a movie throughout this process. The sacrifice in writing this book was mostly felt by them, and I could not have done it without their smiles, love, and support. If I did not have April in my life, my days would be dull, and I wouldn't have such a cheery audience for life's little animations that we share together. She is my best friend forever, the world's greatest cheerleader, a creative and talented Ph.D. candidate, and the model for the person I want to be. I love you all.

About the Author

Brett Feddersen is a certified Project Management Professional (PMP) who holds a masters of professional services in leadership and organizations with an emphasis in strategic innovation from the University of Denver. He is a career public servant having served in the U.S. Marine Corps, and he has worked for the state of Colorado, the city of Boulder (Colorado), and with the Regional Transportation District (RTD) in the Denver/Metro area.

Brett brings a variety of experiences to this book, including project management work on several enterprise resource planning projects, web and e-commerce application implementation, and computer/network infrastructure management. Brett holds the following credentials: Project Management Professional (PMP), Gamification, Lean IT Professional, and ITIL Foundation v3. Brett earned a bachelor's degree in business administration with an emphasis in computer information systems, and he holds a project management certificate from Colorado State University.

It is rare, but not unheard of, to have an extravert in the IT field. Brett would love for you to reach out to him, and you can reach him on LinkedIn: `www.linkedin.com/in/brettfeddersen`.

About the Technical Editor

Vanina Mangano is a program management leader within Google LLC where she leads a team of program managers within the Ads organization. She devotes time to furthering the project management profession through her volunteer work at the Project Management Institute; currently, she serves as an advisory member within PMI's Standards Insight Team, helping to shape the standards roadmap.

Contents

Introduction

Welcome to *CompTIA® Project+® Practice Tests: Exam PK0-005 Second Edition*. Project management remains a fast-growing field that seems to be in constant need of dependable, certified talent. Project managers are needed in just about every field, including, but not limited to, construction, engineering, government and civil initiatives, nonprofits, information technology, logistics, and transportation. Moreover, it can be mighty handy for those projects in your life, including moving, putting in a new kitchen, or helping to organize an event for your local parent-teacher organization. Heck, it was even surprisingly helpful in writing a book of practice questions for the CompTIA Project+ exam.

The Computing Technology Industry Association (CompTIA) promotes the Project+ exam for people interested in project management to earn certifications. The exam helps to validate the various skills, including business and communication skills, that an individual would need to lead projects confidently toward on-time and on-budget completion.

The purpose of this book is to help you pass the CompTIA Project+ Exam PK0-005. As you approach the final validation of your project management knowledge before taking this exam, this book will help prepare you for the types of questions and content that you might encounter on the exam itself. Remember, because of the broad range of knowledge and fields that use project management, this should not be the only resource you use to prepare for the exam. It will be important to use study guides and other exam resources in your studies. The more varied resources to which you expose yourself, the greater the probability of a successful result on the exam.

One such resource you should consider is the *CompTIA Project+ Study Guide: Exam PK0-005 Third Edition* by Kim Heldman (Sybex, 2022). It provides a great overview, reference, and additional review questions as you strive for certification in your project management career.

How This Book Is Organized

This book consists of six chapters, with the first four based on one primarily dedicated to the four domains in the CompTIA Project+ objectives for the PK0-005 exam. The domains for the exam are not evenly weighted with the first two domains, "Project Management Concepts" and "Project Life Cycle Phases" representing 33% and 30% of the examination, respectively. Moreover, those two domains have a much wider scope of material to go over. Accordingly, you will find material that belongs in Domain 1.0 in Chapters 3 and 4. This allows for a complete and proportional covering of all the concepts you will need for the exam and, more importantly, in your project management career. The book also has two chapters that are meant to simulate the exam based on a variety of the questions from all four domains. The chapters are organized as follows:

- Chapter 1, "Project Management Concepts (Domain 1.0)," covers the basic concepts of project management, including project characteristics, comparison of Agile vs. Waterfall project management approaches, change control, schedule development, communications management, effective meeting management, as well as team and resource management.

- Chapter 2, "Project Life Cycle Phases (Domain 2.0)," focuses on the five project phases: Discovery/Concept Preparation, Initiation, Planning, Execution, and Closing.

- Chapter 3, "Tools and Documentation (Domain 3.0)," delves into communication methods, project management productivity tools, quality, and performance to inform project decisions. Additionally, it covers quality management concepts and performance management concepts, and project procurement and vendor selection concepts from Domain 1.0.

- Chapter 4, "Basics of IT and Governance (Domain 4.0)," covers basic environmental, social, and governance (ESG) factors related to project management, reviews information security concepts impacting project management, and explores compliance and privacy considerations. Similar to Chapter 3, this chapter covers a topic from Domain 1.0. This time you will find questions related to risk management.

- Chapter 5 and Chapter 6 each consist of practice exams, representing the mix that you will see on the actual exam from each of the domains. They also carry a representative time limit so that you can simulate the testing experience as a part of your preparation.

The review questions, practice tests, and other testing elements included in this book are *not* derived from the actual Project+ exam questions. They serve to provide exposure to the underlying content and to deliver a comparable testing vehicle for you to prepare for the exam. It will not be a useful exercise to memorize the answers to these questions and assume that doing so will enable you to pass the exam. The underlying subject matter is the important focus of your studies so that you will be able to pass the exam.

CompTIA will rotate through different versions of the exam to help prevent against activities like cheating that would diminish the value of your certifications. Remember that the question bank you have for the exam may be different from that of your peers. As such, this book takes a balanced approach to teaching project management concepts that will generally prepare you to pass any version but may not be able to specifically prep you for a particular, randomized version of the exam. Remember, the ultimate goal of passing the exam is to prepare you for a career in project management. Accordingly, this exam preparation book is an excellent resource to help you with both goals.

How to Use This Book and the Interactive Online Learning Environment and Test Bank

This book includes 1,000 practice test questions, which will help you get ready to pass the Project+ exam. The interactive online learning environment that accompanies the Comp-TIA Project+ practice tests provides a large and varied test bank to help you prepare for the certification exam and increase your chances of passing the first time. There's a tremendous value in taking practice questions as often as possible, even leading up to your actual test. Don't worry if you start to recognize questions from earlier practice runs... that just means you're learning the material and committing it to memory.

The test bank also includes a practice exam. Take the practice exams just as if you were taking the actual exam (without any reference material). As a general rule, you should be consistently making 85% or better before taking the exam.

You can access the CompTIA Project+ Practice Tests Interactive Online Test Bank at www.wiley.com/go/sybextestprep.

Like all exams, the Project+ certification from CompTIA is updated periodically and may eventually be retired or replaced. At some point after CompTIA is no longer offering this exam, the old editions of our books and online tools will be retired. If you have purchased this book after the exam was retired, or are attempting to register in the Sybex online learning environment after the exam was retired, please know that we make no guarantees that this exam's online Sybex tools will be available once the exam is no longer available.

Congratulations on investing in your career and your future as a project manager.

How to Contact the Publisher

If you believe you have found a mistake in this book, please bring it to our attention. At John Wiley & Sons, we understand how important it is to provide our customers with accurate content, but even with our best efforts an error may occur.

In order to submit your possible errata, please email it to our Customer Service Team at wileysupport@wiley.com with the subject line "Possible Book Errata Submission."

Chapter

1

Project Management Concepts (Domain 1.0)

1. What are the defining characteristics of a project? (Choose two.)
   - **A.** A group of related tasks
   - **B.** Temporary in nature
   - **C.** Operational activities
   - **D.** Reworking of an existing project
   - **E.** Creation of a unique product or service

2. The project team is responsible for all of the following, EXCEPT:
   - **A.** Providing governance on the project
   - **B.** Contributing to the deliverables according to the schedule
   - **C.** Contributing expertise to the project
   - **D.** Estimating costs and dependencies

3. A team member is upset about having to stay late as he will miss a scouts meeting with his son. He approaches the project manager, who lets him know that it is a one-time thing and if they work tonight, they will not have to come in this weekend. Upset, the team member stays and finishes his work. This is an example of what type of conflict resolution?
   - **A.** Smoothing
   - **B.** Confronting
   - **C.** Compromising
   - **D.** Avoiding
   - **E.** Forcing

4. When evaluating the project phases, in which phase will project costs be the highest?
   - **A.** Initiating
   - **B.** Discovery
   - **C.** Execution
   - **D.** Closing

5. As a project nears a critical deadline, Ed approaches the project manager and wants to take a couple of days off. The project manager needs Ed's skills to meet the deadline but can see how exhausted Ed has become. They agree that Ed will work through the deadline and then get a couple of days off. Ed accepts this plan and goes back to work. This is an example of what type of conflict resolution?
   - **A.** Smoothing
   - **B.** Confronting
   - **C.** Compromising
   - **D.** Avoiding
   - **E.** Forcing

6. What plan determines the information needs of the stakeholders, format of information delivery, delivery frequency, and the preparer?

    A. Stakeholder Analysis plan

    B. Project Charter

    C. Human Resources plan

    D. Communication plan

7. What is the definition of a portfolio?

    A. A collection of the sample work a project manager has done that they should bring to an interview

    B. A collection of programs, subportfolios, and projects that support strategic business goals or objectives

    C. A group of related projects that are managed together using shared resources and similar techniques

    D. A group of project investments that are maintained to help finance projects

8. In terms of project management, what is a program?

    A. A listing of all individuals involved in the project, including key stakeholders

    B. The software package used to enter and track project management aspects

    C. Related projects that are coordinated and managed with similar techniques

    D. A collection of projects and subportfolios that support strategic goals of the business

9. During an Agile daily stand-up meeting, what are three questions that are asked and answered?

    A. What did I accomplish yesterday? What will I do today? What are the necessary next steps?

    B. What did I accomplish today? Who will I be working with today? What obstacles are preventing progress?

    C. What did I accomplish yesterday? Who will I be working with today? What obstacles are preventing progress?

    D. What did I accomplish yesterday? What will I do today? What obstacles are preventing progress?

10. Marcus works for Wigitcom, and he has been assigned to work on a project. Marcus's regular boss pulls him back to work on assignments and will conduct his performance review. What type of organizational structure is Wigitcom using?

    A. Agile

    B. Functional

    C. Matrix

    D. Projectized

11. A project has a key contributor who is absent from meetings, not meeting deadlines, and affecting the morale of other individuals on a team. There is no other person within the company with the expertise needed to perform the tasks assigned to this team member. Which of the following would be the appropriate action for the project manager?

    A. Bring the team member in for a counseling session.

    B. Leave the employee alone and distribute work to other team members.

    C. Remove the team member from the project and seek a replacement.

    D. Relocate the team member to a different facility.

12. Which of the following is not a role of a project steering committee?

    A. Monitoring project progress

    B. Sprint planning

    C. Advocacy

    D. Offering expert opinion

13. How far in advance should an agenda be published?

    A. 24 hours

    B. 48 hours

    C. 1 week

    D. 1 month

14. A meeting held daily or a couple of times a week where participants give updates on project status, often while standing, is what type of meeting?

    A. Status meeting

    B. Focus group

    C. Stand-up meeting

    D. Demonstration

15. Kim is a project manager who has just been assigned a software development project. She needs help in identifying all the tasks that need to be completed for the project to be successful. What kind of meeting should she use?

    A. Joint application development

    B. Joint application review

    C. Status meeting

    D. Brainstorming

16. In what organizational structure does a project manager have the most limited authority?

    A. Weak-matrix

    B. Projectized

    C. Strong-matrix

    D. Functional

17. A building project requires the following steps: construction, purchasing the build site, blueprinting, and inspection. Purchasing the build site has what relationship to construction?

   A. It is a successor task.

   B. It is a mandatory task.

   C. It is a predecessor task.

   D. It is a discretionary task.

18. Which of the following is the form in which project schedules are typically displayed?

   A. PERT

   B. Calendar

   C. Gantt chart

   D. Pareto chart

19. The project team has been asked to use only the company established tools for instant communication and not use their personal cell phones or computers for such communication. This is an example of:

   A. Communication security

   B. Communication integrity

   C. Communication archiving

   D. Escalating communication issues

20. When the project team is dependent on an entity outside the organization, like a product vendor delivering equipment, this is known as what type of dependency?

   A. Discretionary

   B. Mandatory

   C. External

   D. Financial

21. All electronic communications and meetings notes are stored with the project file at the conclusion of the project. This is an example of:

   A. Communication security

   B. Communication integrity

   C. Communication archiving

   D. Communication planning

22. A project has task A, which will take 2 days; task B, which will take 3 days, task C, which will take 2 days; task D, which will take 2 days; and task E, which will take 3 days. Task A is a predecessor for task B and for task C. Task C is a predecessor for task D. Both task B and task D are predecessors for task E.

   What is the task sequence for the critical path?

   A. A => B => E

   B. A => B => C => D => E

   C. A => B => D => E

   D. A => C => D => E

23. All of the following are aspects of an Agile sprint, EXCEPT:

   A. It is limited to a timeframe such as two weeks.

   B. A planning meeting is held to figure out what the team can accomplish.

   C. A small feature set is taken to completion.

   D. The team works at an incredible pace for the entire time.

24. April works for Wigitcom and has recently been assigned to a project. She was given assignments from both her functional manager and the project manager but was directed to do the project work. When April requested leave, it was granted by her functional manager but then denied by the project manager. What type of organizational structure is Wigitcom using?

   A. Projectized

   B. Strong-matrix

   C. Balanced-matrix

   D. Weak-matrix

25. A meeting to provide a status update on the progress and spending on a project would be what type of meeting?

   A. Collaborative

   B. Informative

   C. Decisive

   D. Focus groups

26. Brad is a junior software developer who is having problems remembering where the source code is for a certain routine. He needs to contact Bridget, the senior developer on the project. What would be the ideal way to contact her?

   A. Email

   B. Impromptu meeting

   C. Instant messaging

   D. Fax

27. A project manager is trying to get a safety update to project team members out in the field. What are appropriate methods to communicate the update to the team? (Choose two).

   A. Fax

   B. Text message

   C. Social media

   D. Voice conferencing

   E. Distribution of printed media

28. New, complex instructions need to be communicated to a project team. What is the best communication method to distribute this information?

   A. Distribution of printed media

   B. Email

   C. Text message

   D. Social media

29. What are the defining characteristics of a project? (Choose two.)

   A. Has a definitive start and end date

   B. Is assigned to a portfolio

   C. Creates a unique product or service

   D. Is a part of ongoing operational activities

   E. Is part of an organization's strategic plan

30. Which of the following describes a portfolio?

   A. A collection of the sample work a project manager has done that should be brought to an interview

   B. A collection of programs, subportfolios, and projects that support strategic business goals or objectives

   C. A group of related projects that are managed together using shared resources and similar techniques

   D. A group of project investments that are maintained to help finance projects

31. DewDrops is a struggling global start-up with a project team located on three different continents. The customer of the project just radically altered one of the triple constraints. What is the best communication method to share this information with the project team?

   A. Impromptu meetings

   B. Virtual meeting

   C. Email

   D. Instant messaging

32. After the project charter is signed, what meeting is held to introduce the project team and stakeholders as well as outlining the goals for the project?

   A. Lessons learned meeting

   B. Project introductory meeting

   C. Kickoff meeting

   D. Team building lunch

33. Which soft skills are important for a project manager?

   A. Time management, earned value calculation, listening, critical path diagrams

   B. Leadership, time management, team building, listening

   C. Time management, earned value calculation, leadership, critical path diagrams

   D. Leadership, following, independence, listening

34. A project is entering the Execution phase, the manager has completed the project planning, and is looking to introduce the key points to the project team. What is the appropriate form of communication for the next step?

  A. Video conference

  B. Impromptu meeting

  C. Email

  D. Kickoff meeting

35. A project schedule serves what function?

  A. Determines the project cost accounting codes

  B. Creates a deliverable-based decomposition of the project

  C. Lists the actions that should be resolved to fulfill deliverables

  D. Determines start and finish dates for project activities

36. Cheryl is a technical lead on a project that is wrapping up remote work at a customer site. What is the best method to communicate the work efforts and next steps with the customer?

  A. Virtual meeting

  B. In-person meeting

  C. Closure meeting

  D. Kickoff meeting

37. All of the following are ways to determine whether a project is completed, EXCEPT:

  A. When the project manager declares the project is complete

  B. When the project is canceled

  C. When it has been determined that the goals and objectives of the project cannot be accomplished

  D. When the objectives are accomplished and stakeholders are satisfied

38. Wigitcom has just been hacked, and millions of records containing personal information of their customers has been stolen. The CEO is in a meeting for the rest of the day. What is the best communication method to let the boss know of the situation?

  A. Social media

  B. Impromptu meeting

  C. Text messaging

  D. Distribution of printed media

39. The creation of a peanut butter and jelly sandwich has the following steps:

Serve.

Gather bread, peanut butter, and jelly.

Place bread on a plate.

Get a knife.

Spread peanut butter on one slice of bread.

Put both slices of bread together.

Spread jelly on the other slice of bread.

What is the correct sequence for this project?

   A. Serve; put both slices of bread together; gather bread, peanut butter, and jelly; get a knife; spread peanut butter on one slice of bread; place bread on a plate; spread jelly on the other slice of bread.

   B. Gather bread, peanut butter. and jelly; get a knife; spread peanut butter on one slice of bread; place bread on a plate; spread jelly on the other slice of bread; serve; put both slices of bread together.

   C. Gather bread, peanut butter, and jelly; get a knife; place bread on a plate; spread peanut butter on one slice of bread; spread jelly on the other slice of bread; put both slices of bread together; serve.

   D. Gather bread, peanut butter, and jelly; get a knife; place bread on a plate; spread peanut butter on one slice of bread; spread jelly on the other slice of bread; serve; put both slices of bread together.

40. Which of the following steps are important in the development of the project schedule? (Choose three.)

   A. Determine tasks.

   B. Set the quality plan.

   C. Set the communication plan.

   D. Sequence the tasks.

   E. Construct a Pareto diagram.

   F. Identify the critical path.

41. The project manager has been gathering a batch of routine organizational messages, project updates, and status reports. What would be the best communication method to share this information?

   A. Video conference

   B. Impromptu meeting

   C. Social media

   D. Scheduled meeting

42. What type of communication method would make sense for routine status meetings on a project where the team is spread out in different cities on the same continent?

   A. In-person meetings

   B. Virtual meetings

   C. Closure meetings

   D. Kickoff meetings

43. To help celebrate the completion of a major project milestone, the project sponsor wants to hold a barbeque on Friday afternoon. What is the best communication method to share this information with the team?

  A. Email

  B. Impromptu meeting

  C. Scheduled meeting

  D. Fax

44. Which of the following are tools and techniques used for developing a project team? (Choose three.)

  A. Team-building activities

  B. Project requirements

  C. Recognition and rewards

  D. Lessons learned meetings

  E. Setting the ground rules

  F. Project kickoff meetings

45. In what stage of team development are teams the most productive and trust levels the highest among team members?

  A. Forming

  B. Storming

  C. Norming

  D. Performing

  E. Adjourning

46. Once a change to the project has been accepted and implemented, what is the next step the project manager should perform?

  A. Conduct a preliminary review.

  B. Collocation.

  C. Determine decision-makers.

  D. Communicate change deployment.

47. A start-up company is attempting to compete in an emerging product market. There are constant disruptive technology changes, and the market is shifting in their product tastes. This type of situation would be best served by which of the following?

  A. Agile approach

  B. Projectized environment

  C. Functional environment

  D. Traditional, or waterfall

48. Kayla works for Wigitcom, and she has been assigned to work on a project. Kayla's project manager gives her direction. At the end of the project, the project manager will conduct Kayla's performance review and she will be free to be given a new work assignment. What type of organizational structure is Wigitcom using?

 A. Agile

 B. Functional

 C. Matrix

 D. Projectized

49. In an Agile methodology, what is a user story?

 A. Key information about stakeholders and their jobs

 B. Short stories about someone using the product or service

 C. Customer survey results after product release

 D. Visual representation of product burndown

50. A large, well-established organization that has been in business for many decades would likely have which organizational structures?

 A. Weak-matrix

 B. Projectized

 C. Strong-matrix

 D. Functional

51. Marcus is working on a project to build a new video game. His boss asks him what it would take to make a major change to the platform engine. What would be Marcus's next step?

 A. Validate the change implementation.

 B. Implement changes.

 C. Conduct an impact assessment.

 D. Escalate to the CCB.

52. Identified in the form of user stories, what is a list of all things to be completed, whether technical or user-centric in nature, known as?

 A. Requirements

 B. Backlog

 C. Risk register

 D. Stakeholders

53. Thaala is a project manager creating an agenda for an upcoming meeting. It must be made for the project to stay on track. She gives 15 minutes on the agenda for each of the six items that must have direction for meeting to be successful. What activity has Thaala performed?

 A. Timeboxing

 B. Sprint planning

 C. Creating action items

 D. Joint application development

54. What are milestones?

    A. A measure of the distance traveled on a project

    B. Characteristics of deliverables that must be met

    C. Checkpoints on a project to determine Go/No-Go decisions

    D. Major events in a project used to measure progress

55. The DewDrops project team is meeting with product stakeholders to go over recent product builds to ensure it meets expectations and requirements. What type of meeting is this?

    A. Joint application development session

    B. Joint application review session

    C. Joint budget committee

    D. Project steering committee

56. Which meeting role is responsible for leading the meeting by ensuring things run smoothly and that all goals of the meeting have been met by its end?

    A. Facilitator

    B. Scribe

    C. Timekeeper

    D. Coordinator

57. All of the following are types of dependencies, EXCEPT:

    A. Mandatory

    B. Discretionary

    C. External

    D. Backlog

58. A stakeholder has asked to add a feature to a project, but the request is rejected by the project manager. What is the likely reason the scope was rejected?

    A. There is interaction between constraints.

    B. Scope creep is occurring on the project.

    C. The request can be handled without the formality.

    D. The sponsor is on vacation.

59. Creating a list of items that will be covered during a meeting is called:

    A. Task setting

    B. Action items

    C. Follow ups

    D. Agenda setting

60. The project needs time for the team to work on the creation of an artifact. Which meeting type would be best for the team to use?

   A. Stand-up

   B. Focus groups

   C. Presentation

   D. Workshop

61. A company is expanding and has several projects underway. One project is building a new wing on the headquarters building, and the other is installing a new high-speed fiber network. The framing of the new building must begin before the installation of the new network can begin. This is an example of what type of logical relationship?

   A. Finish-to-finish

   B. Start-to-start

   C. Finish-to-start

   D. Start-to-finish

62. A project has task A, which will take 2 days; task B which will take 3 days; task C, which will take 2 days; task D, which will take 2 days; and task E, which will take 3 days. Task A is a predecessor for task B and for task C. Task C is a predecessor for task D. Both task B and task D are predecessors for task E.

   What is the duration of the critical path?

   A. 7 days

   B. 9 days

   C. 10 days

   D. 11 days

63. In what project artifact would you find information relating to the quality and availability of resources?

   A. Project schedule

   B. Organization chart

   C. Resource calendar

   D. Risk register

64. DewDrops has a project to implement a new human resource information system (HRIS) for the company. They have changed their minds during implementation, and they want to add an additional module to the final product. What type of change is this?

   A. Project change

   B. Organizational change

   C. Product change

   D. Mindset change

65. At the end of a meeting, any assignments that were made, who was given that assignment, and the expected due date for completion are reviewed. This is a review of:

    A. Follow-ups

    B. Refinement

    C. Brainstorming

    D. Action items

66. Nyssa works for Wigitcom and has been assigned to a project. She wants to take a week off for a family reunion, which conflicts with a project deadline. The project manager denied her leave request, but her functional manager lets her attend the family reunion. What type of organizational structure is Wigitcom using?

    A. Projectized

    B. Strong-matrix

    C. Balanced-matrix

    D. Weak-matrix

67. When is a project considered to be a success?

    A. Stakeholder expectations have been met.

    B. The phase completion has been approved.

    C. All project phases have been completed.

    D. The vendor has been released from the project.

68. Which of the following in an example of a deliverable?

    A. The date work on the project begins

    B. The design for a new product

    C. Time and materials applied to the project

    D. PMO

69. A project sponsor would be responsible for all of the following, EXCEPT:

    A. Developing high-level requirements for the project

    B. Functioning as the approval authority and removing roadblocks

    C. Marketing the project across the organization

    D. Estimating the costs and dependencies of the project activities

    E. Serving to help control the direction of the project

70. A building project requires the following steps: construction, purchasing the build site, blueprinting, and inspection. Construction has what relationship to blueprinting?

    A. It is a successor task.

    B. It is a mandatory task.

    C. It is a predecessor task.

    D. It is a discretionary task.

71. All projects are constrained by which three elements as they affect quality?

  A. Time, budget, scope

  B. Time, risks, budget

  C. Cost, benefits, scope

  D. Cost, risks, scope

72. What key milestone is triggered when the project charter is signed?

  A. A project sponsor can now be chosen.

  B. Key stakeholders are freed from project communication.

  C. The project is authorized to begin.

  D. Project resources are released from the project.

73. Amy is a project manager and assigns Oswald to capture the meeting minutes. What role did she assign Oswald?

  A. Facilitator

  B. Scribe

  C. Timekeeper

  D. Coordinator

74. All of the following are examples of project resources, EXCEPT:

  A. Team members

  B. Equipment

  C. WBS

  D. Materials

75. What is resource smoothing?

  A. Accommodating resource availability within activity float times

  B. A schedule compression technique

  C. An attempt to balance assignments to prevent overload

  D. A method for loading heavy equipment

76. What is resource leveling?

  A. A storage technique for physical resources

  B. A schedule compression technique

  C. An attempt to balance assignments to prevent overload

  D. A method for loading heavy equipment

77. Fast-tracking a project is a technique involving which of the following?

  A. Performing two tasks in parallel that were previously scheduled to start sequentially

  B. Looking at cost and schedule trade-offs such as adding more resources

  C. Moving later deliverables to earlier phases to appease stakeholders

  D. Removing critical path activities that are unnecessary

78. The longest full path of any project is known as:
   A. Critical path
   B. Total float
   C. Delphi technique
   D. Pareto analysis

79. What does a resource shortage mean?
   A. There is a shortage of things for team members to work on.
   B. Not enough resources are available for the task, leading to overallocation.
   C. There are too many resources, leading to underallocation.
   D. There is an abundancy of things for team members to work on.

80. In what stage of team development do the members stop working with one another and return to their functional jobs?
   A. Forming
   B. Storming
   C. Norming
   D. Performing
   E. Adjourning

81. Kayla is a project manager and has received a request to move up the due date of a project. How should she keep track of this change?
   A. Document the request in the change control log.
   B. Update the scope statement.
   C. Place the change on the Kanban board.
   D. Update this in the requestor's personal file.

82. WigitCom completed a project last year to roll out a new asset management platform of the organization. They are now looking to add an additional module to the platform. What type of change is this?
   A. Project change
   B. Organizational change
   C. Product change
   D. Asset change

83. An organization that has a fixed budget and offers a stable environment would be best suited for which type of project management approach?
   A. Traditional, or waterfall
   B. Projectized environment
   C. Agile approach
   D. Functional environment

84. All of the following are life cycle phases of a project, EXCEPT:

    A. Planning

    B. Closing

    C. Development

    D. Execution

85. What does a change control board (CCB) do to support the project?

    A. Helps vet and manage changes to the scope

    B. Provides an accounting structure for tasks

    C. Sets the standards and templates for the project

    D. Sets the costs of quality for the project

86. Which project role is responsible for coordinating resources between projects?

    A. Project management office (PMO)

    B. Project coordinator

    C. Project manager

    D. Project scheduler

87. The CCB has approved a change to the project that will add $5,000 to the cost and an additional week to the project duration. What should be the next step that happens?

    A. Escalate the decision to the executive committee.

    B. Consult the communication plan and inform the appropriate stakeholders.

    C. Determine the organization's tolerance for this change.

    D. Hold a Scrum meeting to see what to work on next.

88. There is a difference of opinion between a group of stakeholders and the project team regarding the need of a change to the agreed upon requirements. They need help in making a decision; who should this be escalated to for a final decision?

    A. Project sponsor

    B. Project manager

    C. Executive steering committee

    D. Change control board

89. When should employee performance expectations be set on a project?

    A. Lessons learned meeting

    B. Employee performance review

    C. Stakeholder identification meeting

    D. First meeting with a new team member

90. All of the following are used in an Agile approach to project management, EXCEPT:

  A. Burndown charts

  B. WBS

  C. Continuous requirements gathering

  D. Sprint planning

91. A project team is using a series of meetings where the software developers, business analysts, and product stakeholders meet frequently to collect and analyze software requirements. What type of meetings are these?

  A. Stand-ups

  B. Brainstorming

  C. Joint application review sessions

  D. Joint application development sessions

92. Teams normally go through a similar development cycle. Which is the correct order of those stages?

  A. Norming, Forming, Storming, Adjourning, Performing

  B. Forming, Storming, Norming, Performing, Adjourning

  C. Forming, Norming, Performing, Storming, Adjourning

  D. Norming, Storming, Forming, Adjourning, Performing

93. In the development of a project schedule, the need to set governance gates is important. All of the following are examples of governance gates, EXCEPT:

  A. Daily stand-up meetings

  B. Client sign-off

  C. Management approval

  D. Legislative approval

94. Which project role outlines the consequences of nonperformance?

  A. Project manager

  B. Project coordinator

  C. Project scheduler

  D. Project management office

95. WigitCom has undertaken a project to build a new app for the medical community. It holds a meeting to demonstrate its project prototype to doctors, nurses, administrators, and medical boards. How would you categorize this group of medical attendees in the context of the meeting?

  A. Steering committee members

  B. Shareholders

  C. Target audience

  D. Business analysts

96. The types of organizational structures include which of the following? (Choose three.)

- **A.** Agile
- **B.** Functional
- **C.** Matrix
- **D.** Colocation
- **E.** Projectized

97. A meeting was convened with project decision-makers to discuss new developments in the funding for the project and whether to change the scope of the project or potentially seek other alternatives to funding. What type of meeting is this?

- **A.** Stand-up
- **B.** Workshop
- **C.** Steering committee
- **D.** Demonstration

98. DewDrops is about to launch a new software product to the market. They are concerned about a new regulation that is being considered in a large state that may cause modification to the design, specifically around data privacy. It was suggested during a steering committee meeting that this issue get revisited after the final vote on the new law. What is this an example of?

- **A.** Follow-up
- **B.** Action item
- **C.** Task setting
- **D.** Refinement

99. A project management office (PMO) has which of the following responsibilities? (Choose three.)

- **A.** Markets the project across the business
- **B.** Provides governance for projects
- **C.** Manages the team, communication, scope, risk, budget, and time of the project
- **D.** Maintains standard documentation and templates
- **E.** Establishes key performance indicators and parameters
- **F.** Develops and maintains the project schedule

100. Which project role helps to market the need and success of the project and provides a level of control for funding?

- **A.** Project sponsor or champion
- **B.** Project manager
- **C.** Project coordinator
- **D.** Project scheduler

101. In terms of project management, what is a program?

    A. A listing of all individuals involved in the project, including key stakeholders

    B. The software package used to enter and track project management aspects

    C. Related projects that are coordinated and managed with similar techniques

    D. A collection of projects and subportfolios that support the strategic goals of the business

102. Project manager Allison was asked by a project stakeholder why a key piece of functionality on a software project was struck from the scope list. Allision remembers a discussion that took place during a steering committee meeting. Which artifact should Allison review to find out more information on this decision?

    A. Project charter

    B. Ask the project sponsor

    C. Meeting minutes

    D. Action items

103. The project champion is eager to see the new product prototype in action. Which is the best meeting format to share this information?

    A. Status meeting

    B. Demonstration

    C. Brainstorming

    D. Focus group

104. The project team needs to gather the opinions of the product users from finance on requirements for the new ERP system. What is the appropriate type of meeting?

    A. Stand-up

    B. Focus groups

    C. Presentation

    D. Workshop

105. Progressively elaborating deliverables into differing levels of a WBS is known as:

    A. Producing a backlog

    B. Progressive iteration

    C. Rolling wave planning

    D. Prioritizing tasks

106. While sequencing activities, a project manager notices an activity that cannot begin until a different task is completed. Which of the logical relationships is the project manager most likely to use?

    A. Start-to-start

    B. Finish-to-start

    C. Start-to-finish

    D. Finish-to-finish

107. In what stage of team development do team members begin to confront each other and vie for position and control?

   A. Forming

   B. Storming

   C. Norming

   D. Performing

   E. Adjourning

108. Which of the following is a simple time management technique where a fixed maximum amount of time for an activity is set in advance, and then the activity is completed within that timeframe?

   A. Sprint planning

   B. Parametric estimating

   C. Timeboxing

   D. Scrum

109. When a meeting agenda sets a fixed time on a schedule where a task must be completed within that time, what technique is being used?

   A. Soft timeboxing

   B. Action items

   C. Brainstorming

   D. Hard timeboxing

110. In an Agile approach to project management, what is a backlog?

   A. Delayed work that is caused by bottlenecks

   B. Customer prioritized functionality list that still needs to be added to the product

   C. Inventory not added to the project because of shipping delays

   D. A daily meeting focusing on three questions

111. When estimating activity resources, a project manager looks to variations and other options to complete the work. What is this tool or technique called?

   A. Parametric estimating

   B. Bottom-up estimating

   C. Alternative analysis

   D. Fishbone diagraming

112. Nestor is a project manager assigned to build a new branch office for a bank. The branch office will be of a similar size and design as another branch office. He has been asked to create cost and schedule estimates and to follow the bank's best practices for projects. Whose responsibility is it to help Nestor with this effort?

   A. Project sponsor

   B. Project management office

   C. Project team

   D. Project scheduler

113. Crashing is a project management technique involving:

   A.   Performing two tasks in parallel that were previously scheduled to start sequentially

   B.   Looking at cost and schedule trade-offs like adding more resources

   C.   Moving later deliverables to earlier phases to appease stakeholders

   D.   Removing critical path activities that are unnecessary

114. Fast tracking a project is a technique involving:

   A.   Performing two tasks in parallel that were previously scheduled to start sequentially

   B.   Looking at cost and schedule trade-offs like adding more resources

   C.   Moving later deliverables to earlier phases to appease stakeholders

   D.   Removing critical path activities that are unnecessary

115. All of the following techniques can be used to estimate the duration of an activity, EXCEPT:

   A.   Expert judgment

   B.   Three-point estimating

   C.   Analogous estimating

   D.   Pareto diagramming

116. When there are scarce resources to perform specific activities on a project, and the activities must be completed at certain times, which tool or technique would you use?

   A.   Fast tracking

   B.   Reverse resource allocation

   C.   Resource smoothing

   D.   Resource leveling

117. The practice of adding a percentage of time to a work package, or adding a percentage of money to a project as an emergency, is known as:

   A.   Risk response plan

   B.   Contingency reserves

   C.   Allowing risk strategy

   D.   Delphi technique

118. Which of the following are true regarding kickoff meetings?

   A.   First time project team members are introduced to one another.

   B.   Happens in the Initiation phase.

   C.   Occurs when project planning is complete.

   D.   All of the above.

   E.   A and C.

119. Project managers should spend how much of their time communicating?

    A. Up to 40%

    B. Up to 50%

    C. Up to 75%

    D. Up to 90%

120. A project stakeholder has which of the following responsibilities?

    A. Documentation and administrative support, estimation of task duration, soliciting task status from resources, expertise

    B. Vested interest, providing input and requirements, project steering, expertise

    C. Documentation and administrative support, providing input and requirements, project steering, expertise

    D. Vested interest, providing input and requirements, cross-functional coordination, expertise

121. Jenny works for a company undertaking a project. She will ultimately benefit from the service created and would like to share her thoughts and input on how it should be created. She is also a subject matter expert in the product area. Jenny is most likely which of the following?

    A. Project stakeholder

    B. Project champion

    C. Project sponsor

    D. Member of the PMO

122. What are two types of discretionary fund allocations that a project may be granted?

    A. Top-down and bottom-up

    B. Contingency and discretionary

    C. Parametric and analogous

    D. Contingency and management

123. DewDrops has wrapped up project work, and they are looking to hand off the project to the customer and revisit how the project went for everyone. What is the appropriate communication method to conduct these activities?

    A. Kickoff meeting

    B. Virtual meeting

    C. Email

    D. Closure meeting

124. Pete is a human resources manager who needs to communicate a complaint issued against Ashley. How would the content of this message dictate what type of communication is used with Ashley?

    A. It is of a confidential nature so meet with her face-to-face.

    B. Personal preferences suggest using social media.

    C. Voice conferencing avoids personal threat.

    D. Instant messaging allows time to digest information slowly.

125. Which project role is responsible for all project artifacts like project plans, meeting minutes, and project delivery?

    A. Project coordinator

    B. Scheduler

    C. Project team

    D. Project manager

126. What are project requirements?

    A. A measure of the distance traveled on a project

    B. Characteristics of deliverables that must be met

    C. Checkpoints on a project to determine Go/No-Go

    D. Major events in a project used to measure progress

127. When using a waterfall methodology, what are critical elements that need to be included in the project schedule?

    A. Define activities, sequence activities, estimate resources, estimate duration

    B. Define activities, budget activities, estimate resources, estimate completion

    C. Budget activities, estimate resources, determine milestones, estimate completion

    D. Develop schedule, determine completion date, check stakeholder assumptions, conduct feasibility assessment

128. Nora is a project manager implementing an enterprise software package and she has been assigned an onshore and an offshore team. What are factors that will influence communication methods on the project? (Choose three.)

    A. Time zones

    B. Cultural differences

    C. Level of report detail

    D. Criticality factors

    E. Language barriers

    F. Technological factors

129. Murthy is a project manager working on an international software project. What is a factor influencing the communication methods on the project?

    A. Language barriers

    B. Fax

    C. Risk identification

    D. Scope creep

130. Which of the following is the sprint planning meeting used to do?

    A. Get a head start on the work needed for the project.

    B. Prepare the project charter and kickoff meeting.

    C. Set a realistic backlog of items completed during this iteration.

    D. Set the communication and quality plans for the project.

131. Gus is an oil field team member assigned to a project in a remote area. The trailer he works out of has a telephone line, but internet connectivity and cell phone coverage are spotty. What would be the best method to send Gus's instructions for the next day?

    A. Email

    B. Impromptu meeting

    C. Instant messaging

    D. Fax

132. Phil and Fernando are technicians working on a project out in the field. Power was supposed to be turned off at 5 p.m. so work can begin, but there is going to be a delay. What is the best way to communicate this change?

    A. Email

    B. Phone call

    C. Text message

    D. Scheduled meeting

133. Mandy is a project manager for a team located at three locations around the globe. She just received an update from the customer of an urgent nature. What type of communication method should Mandy use to get this information out to the team?

    A. Email

    B. Impromptu meeting

    C. Virtual meeting

    D. Scheduled meeting

134. Which of the following factors present challenges for a global project team located on different continents? (Choose three.)

    A. Language barriers

    B. Level of report detail

    C. Technological factors

    D. Time zones

    E. Cultural differences

    F. Criticality factors

135. When considering the basic communication model, what are the basic elements that are needed? (Choose three.)

    A. Transmission

    B. Receiver

    C. Inbox

    D. Sender

    E. Nonverbal communication

    F. Decoder

    G. Message

136. All of the following are common types of project changes EXCEPT:

    A. Scope changes

    B. Requirement changes

    C. Organizational changes

    D. Timeline changes

137. Once a change request is submitted, where should it be recorded and assigned an identification number for tracking purposes?

    A. Change request log

    B. Risk register

    C. Business process repository

    D. Issue log

138. Which document will ensure the capture of all needed change request information so that it can receive proper consideration?

    A. SIPOC-R

    B. Project charter

    C. Template

    D. Meeting minutes

139. In what type of organizational structure would resources report solely to the project manager?

    A. Weak-matrix

    B. Projectized

    C. Strong-matrix

    D. Functional

140. Predecessor and successor tasks can have four possible logical relationships. Which of the following is not one of them?

   A. Finish-to-finish

   B. Start-to-deferred

   C. Finish-to-start

   D. Start-to-finish

141. What aspect of project management is shared with Agile and other approaches?

   A. Sprint planning

   B. Self-organized and self-directed teams

   C. Iterative approach

   D. Adaptive to new/changing requirements

142. Before a construction company can begin building a road over a mountain pass, they must wait for the spring thaw to occur so they can get heavy machinery into the work location. This is known as what type of dependency?

   A. Internal

   B. Discretionary

   C. External

   D. Mandatory

143. A software development company is considering using an Agile approach to a new project. They would use all of the following EXCEPT:

   A. Gate checks

   B. Burndown charts

   C. Sprint planning

   D. Continuous requirements gathering

144. In terms of resource assignments, which best describes how resources are assigned in a projectized environment?

   A. Resources are assigned on an ad hoc basis.

   B. Resources are assigned from a functional area to the project.

   C. Resources must be outsourced.

   D. Resources must not be collocated.

145. Benched resources are:

   A. Great for an organization as there is always staff to work on a project

   B. Bad for an organization as there is always staff to work on a project

   C. Costly because individuals are being paid to sit around

   D. Inexpensive because individuals are not being paid when they sit around

146. When would an adaptive method be preferable to a more rigid project management style?

    A. In a mature organization with defined processes

    B. When the scope can be easily and thoroughly defined

    C. Where small incremental improvements offer no value to stakeholders

    D. When an organization is dealing with a rapidly changing environment

147. The Transportation Exchange is a new ride sharing service trying to break into the marketplace. Which of the following benefits would make sense for them to use an adaptative method of project management?

    A. There is no value with small incremental improvements.

    B. Processes are thoroughly defined.

    C. The environment is rapidly changing.

    D. Scope is easily identified and designed.

148. The Transportation Exchange project team is assembled, and introductions are being performed. What stage of team development is this?

    A. Storming

    B. Norming

    C. Performing

    D. Forming

149. Adric was recently assigned to a project at WigitCom. He received task assignments from both the project manager and his normal supervisor. He is notified that both the project manager and the supervisor will contribute to his performance review. What type of organizational structure is WigitCom using?

    A. Projectized

    B. Strong-matrix

    C. Balanced-matrix

    D. Weak-matrix

150. WigitCom has a mobile geolocation application that was released last year. They are now working on the latest quarterly release of the application, which has minor updates and bug fixes. Which of the following statements is true regarding the geolocation application effort? (Choose all that apply.)

    A. This is a project because there are minor changes to the application.

    B. This is not a project because the regular releases are a continuing effort.

    C. This is a project because this effort is temporary in nature.

    D. This is not a project because the product being produced is not unique.

151. An iterative, incremental approach to managing the activities on a project in a highly flexible manner is referred to as:

   A. Waterfall methodology

   B. Matrixed management

   C. Projectized

   D. Agile methodology

152. A project has a team member who is absent from meetings, is not meeting deadlines, and is affecting the morale of other individuals on a team. The appropriate action for the project manager would be which of the following?

   A. Bring the team member in for a counseling session.

   B. Leave the employee alone and distribute work to other team members.

   C. Remove the team member from the project and seek a replacement.

   D. Relocate the team member to a different facility.

153. What are the defining characteristics of a project? (Choose two.)

   A. An organized effort to fulfill a purpose

   B. Blueprints needed to construct a building

   C. Routine activities to an organization

   D. Has a specific end date

   E. Reworking an existing project

154. Ashley is a program manager for the construction of several transit projects. She asks the bridge project manager for updated estimates on the bridge's construction. Who has the responsibility for the estimating task duration and costs?

   A. Project manager

   B. PMO

   C. Stakeholders

   D. Project team

155. Not including time off, holidays, or nonproject work, the total time involved for an individual to complete a task is:

   A. Analogous estimating

   B. SWAG

   C. Work effort estimate

   D. SPI

156. In a situation where the end product is uncertain and/or the conditions for developing a product or service are in flux, what would be the best project management approach?

   A. Traditional, or waterfall

   B. Projectized environment

   C. Agile approach

   D. Functional environment

157. Tiffany is the only digital marketer assigned to the project, but only 60 percent of her time is available to the project. There is enough work for a person assigned 100 percent of the time to the project, so Tiffany is struggling to meet her deadlines. This is an example of which of the following?

    A. Low-quality resources

    B. Interdependencies

    C. Dedicated resources

    D. Resource overallocation

158. Amy has been assigned to a project and reports to Kim, the project manager. John, Amy's functional manager, also requires Amy to report to him. What type of resource is Amy?

    A. Dedicated

    B. Physical

    C. Digital

    D. Shared

159. Obtaining a sign-off on the design of a product would be an example of which type of dependency?

    A. Discretionary

    B. Mandatory

    C. External

    D. Financial

160. A project manager is looking to boost the morale of the team through a meeting that includes both social and business aspects. The project manager is engaged in what type of activity?

    A. Trust building

    B. Forming

    C. Team building

    D. Management skills

161. A visual representation of how quickly requirements are being completed with each iteration is called:

    A. Fishbone diagram

    B. Burndown chart

    C. Gantt chart

    D. Pareto diagram

162. WigitCom is faced with changes to scope and personnel on a project. What is the appropriate method to share this information with stakeholders?

    A. Social media and text messages.

    B. Memo and email.

    C. Hold a meeting with the project team.

    D. Let the communication plan be the guide.

163. At the completion of a project sprint, the project team meets to examine what went well, what didn't go well, and what improvements could be made. This is an example of which of the following?

A. Governance gates

B. Product backlog

C. Daily Scrum

D. Scrum retrospective

164. WigitCom has a group of projects all related to security widgets. They are wanting to add a new product for security cameras and sell them to customers. The effort must be completed within the next three months to beat the competition to market. There is a group of resources who work on security efforts. Which of the following are true about this effort? (Choose three.)

A. This effort is not a project because security is already done.

B. This effort is a project because the product being developed is unique.

C. This effort is a project and will be a part of a program.

D. There is no reason to do this effort because they already do security.

E. This meets the requirements for a project because it creates a unique product and is temporary in nature.

165. With an Agile methodology, all of the following are true with an adaptive life cycle, EXCEPT:

A. Requires a high degree of stakeholder involvement.

B. All requirements must be gathered up front.

C. Rapid iterations.

D. Fixed time and resources.

166. The development of architectural design took twice the amount of time as was projected. What type of project change would this represent?

A. Scoping change

B. Timeline change

C. Quality change

D. Funding change

167. After a productive week, a project team in the field has lots of detailed updates to share with the project manager back at headquarters. The work lead spends an hour writing up all the accomplishments, problems, and next steps for the field team. What factors influence this communication?

A. Tailor method based on content of message

B. Intraorganizational differences

C. Personal preferences

D. Criticality factors

168. After establishing the product backlog, what tool would be used to determine the project's velocity?

    A. Fishbone diagram

    B. Burndown chart

    C. Gantt chart

    D. Pareto diagram

169. After a subject matter expert (SME) evaluates the impacts of a change, they should then analyze the following specific elements of the change EXCEPT:

    A. Additional equipment needs

    B. Costs

    C. Approve or reject the change

    D. Resource hours needed

170. After a change request has been recorded in the change request log, what is the next step that should be performed?

    A. Submit to CCB to be accepted or rejected.

    B. Defer until there is a break in the project schedule.

    C. Implement the change.

    D. Analyze the impact of the change.

171. In what stage of team development do things begin to calm down because the team members become more comfortable with one another?

    A. Forming

    B. Storming

    C. Norming

    D. Performing

    E. Adjourning

172. A change control template should include all of the following EXCEPT:

    A. The change that is requested

    B. The reason for the change

    C. The executive sponsor of the change

    D. What will happen if the change is not made

173. What is the appropriate method for submitting a change request?

    A. In writing

    B. Verbally

    C. Via videoconference

    D. Executive session of the change control board

174. As a project manager, a dedicated resource would be the ideal situation because:

  A. The team member will continue to share time with their functional work.

  B. The project manager has full authority and controls time and tasks.

  C. The dedicated resource won't have to be paid overtime.

  D. Low-quality resources aren't assigned to a project.

175. When a dependency is directly related to the type of work on which it is being performed, it is what type of dependency?

  A. Discretionary

  B. Mandatory

  C. External

  D. Financial

176. Who can submit a change request?

  A. Project manager

  B. Almost anyone working on or associated with the project

  C. Project sponsor

  D. Change control board

177. After a project change is identified, evaluated, and approved, what is the next step in the change control process?

  A. Update documents.

  B. Validate the change; do a quality check.

  C. Obtain approval.

  D. Implement change.

178. Avinash is a senior database administrator assigned to a project. After two months, he hands off the work to Alyson, who is a journey-level database administrator. This represents what type of common project change?

  A. Timeline change

  B. Requirements change

  C. Quality change

  D. Resource change

179. During a merger, the new company has decided to expand their headquarters building, which was a project currently in construction. They have allocated an additional $300,000 to the project to make this happen. Which common project changes does this represent? (Choose two.)

  A. Requirements change

  B. Scoping change

  C. Timeline change

  D. Funding change

180. Wigitcom is faced with changes to scope and personnel on a project. What is the appropriate method to share this information with the stakeholders?

A. Via social media and text messages

B. Using memos and email

C. Holding a meeting with the project team

D. Following the communication plan

181. Marion is a project manager working on implementing a new asset management system for an agency. She has encountered problems when trying to get participation from other departments, and it is creating problems. Whose responsibility would it be to help clear the obstacle?

A. Project coordinator

B. PMO

C. Stakeholders

D. Project sponsor

182. Wigit Construction's customer has changed their mind on the color of the exterior and has decided that the bathrooms in a building all need to be accessible to those with disabilities. This is an example of what type of project change?

A. Resource change

B. Requirements change

C. Funding change

D. Scope change

183. WigitCom's customer has asked for a reporting engine to be added to their software, which was not originally part of the scope. Which common project changes does this represent? (Choose two.)

A. Scoping change

B. Funding change

C. Quality change

D. Timeline change

184. A project has fallen behind schedule, and the project manager has decided to run the next several activities in parallel instead of sequentially to help make up time. This is an example of:

A. Crashing

B. Risk avoidance

C. Fast tracking

D. Critical path method

185. Wigit Construction is organized by projects where the project managers have ultimate authority over resources like personnel and equipment. Which type of organization is Wigit Construction?

A. Projectized

B. Matrix

C. Functional

D. Agile

186. What is rolling wave planning?

A. Planning for areas of intense activity to allocate team members according to the resource plan

B. The process of progressively elaborating deliverables or project phases into differing levels of the WBS

C. A design technique used to ensure the structural integrity for earthquake-proof buildings

D. A quick-start technique of where to begin a project with little planning or sign-off to generate momentum

187. Judy is a program manager and is monitoring the work done on several projects. On the telecom project, she needs more information on when certain activities and milestones will occur. Who on the telecom project should Judy reach out to for this information?

A. Project scheduler

B. Project coordinator

C. PMO

D. Project manager

188. Madeline serves on a project, is a liaison between senior management and the project team, maintains team management and quality control, and has good knowledge of Agile, Scrum and or Kanban. What is Madeline's role on the project?

A. Business analyst

B. Program manager

C. Scrum master

D. Developers/engineers

189. A project is underway, and the team has missed several deliverable dates. The steering committee would like the project to stay on track, and the project manager indicates that the team will need to work overtime to make the deadline. The increased cost is restricted by which type of constraint?

A. Environment

B. Budget

C. Scope

D. Scheduling

190. Which project role supervises projects and teams with a focus on the successful launch of a product?

   A. Product manager

   B. Project manager

   C. Senior management

   D. Product owner

191. The scope baseline allows project managers to perform all of the following activities, EXCEPT:

   A. Set the approach to conflict resolution.

   B. Document schedules.

   C. Assign resources.

   D. Monitor and control project work.

192. In a situation where the end product is uncertain and/or the conditions for developing a product or service are in flux, what would be the best project management approach?

   A. Traditional, or waterfall

   B. Projectized environment

   C. Agile approach

   D. Functional environment

193. All of the following are characteristics of an Agile project management approach, EXCEPT:

   A. Strict adherence to a change control process.

   B. Uses a flexible approach to requirements.

   C. Team members work in short bursts, or sprints.

   D. Each release is tested against the customers' needs.

194. Which of the following is not a principle of extreme programming?

   A. Rapid feedback

   B. Embracing change

   C. Project phases do not overlap

   D. Assume simplicity

195. The sprint planning meeting is used to achieve which of the following?

   A. Getting a head start on the work needed for the project

   B. Preparing the project charter and kickoff meeting

   C. Setting a realistic backlog of items completed during this iteration

   D. Establishing the communication and quality plans for the project

196. Which of the following are characteristics of an Agile project management approach? (Choose three.)

    A. Self-organized teams

    B. Sprint planning

    C. Upfront, comprehensive requirements gathering

    D. Formally organized teams

    E. Continuous requirements gathering

    F. Feedback based primarily in lessons learned meetings

197. Which methodology or framework is the union of people, process, and technology to continually provide value to customers?

    A. Scaled Agile Framework (SAFe)

    B. Waterfall

    C. DevOps

    D. Scrum

198. What are four of the responsibilities of a project sponsor? (Choose four.)

    A. Develops the business case and justification

    B. Functions as the approval authority for funding

    C. Sets the standards and practices a project

    D. Provides input and requirements

    E. Helps to control the project's direction

    F. Manages the risks of the project

199. What are the five main components of the Scaled Agile Framework (SAFe)?

    A. Requirements, Design, Implementation, Verification/Testing, Deployment/Maintenance

    B. Requirements, Sprint Planning Meeting, Discovery, Testing, Closing

    C. Architecture, Integration, Governance, Funding, Roles

    D. Discovery, Sprint Planning Meeting, Testing, Funding, Closing

200. Which type of project management certification provides a methodology to perform and complete the project?

    A. Project Management Professional (PMP)

    B. Software Development Life Cycle (SDLC)

    C. Extreme Programming (XP)

    D. PRojects IN Controlled Environments (PRINCE2)

201. Which methodology focuses on just-in-time delivery of functionality and managing the amount of work in progress (WIP)?

   A. Waterfall

   B. Kanban

   C. DevOps

   D. SDLC

202. Which methodology would be used to build an online load application for a finance start-up?

   A. Kanban

   B. Agile

   C. DevOps

   D. DevSecOps

203. Which project management methodology follows a chronological process and works based on fixed dates, requirements, and outcomes?

   A. DevOps

   B. Waterfall

   C. Scrum

   D. Scaled Agile Framework

204. A project team is assigned two individuals directly out of college with no experience working in advanced electronics. The two team members cannot be assigned work without a more senior team member working alongside them. This is an example of which one of the following?

   A. Shared resources

   B. Resource shortage

   C. Low-quality resources

   D. Benched resources

205. What are the five common stages in a Waterfall process?

   A. Requirements, Design, Implementation, Verification/Testing, Deployment/Maintenance

   B. Requirements, Sprint Planning Meeting, Discovery, Testing, Closing

   C. Architecture, Integration, Governance, Funding, Roles

   D. Discovery, Sprint Planning Meeting, Testing, Funding, Closing

Chapter
2

Project Life Cycle
Phases (Domain 2.0)

1. What elements are explained in a business case?
    A. Justification by identifying the organizational benefits
    B. Alternative solutions
    C. Alignment to the strategic plan
    D. All of the above
    E. A, C

2. What plan determines the information needs of the stakeholders, the format of information delivery, delivery frequency, and the preparer?
    A. Stakeholder analysis plan
    B. Project charter
    C. Human resources plan
    D. Communication plan

3. The high-level scope definition describes which of the following?
    A. High-level deliverables of the project
    B. Objectives of the project
    C. Reason for the project
    D. All of the above

4. In what project life cycle phase are the majority of the processes and project documents created?
    A. Discovery/concept preparation
    B. Initiation
    C. Planning
    D. Execution
    E. Closing

5. Which component of the project charter describes the characteristics of the product created by the project?
    A. Project description
    B. Business case
    C. Deliverables
    D. Quality plan

6. In what project life cycle phase is the influence of stakeholders the least effective?
    A. Discovery/concept preparation
    B. Initiation
    C. Planning
    D. Execution
    E. Closing

7. Which conflict resolution technique produces a win-lose result for the parties?

   A. Forcing

   B. Confronting

   C. Avoiding

   D. Attacking

8. What are the three common constraints found in projects? (Choose three.)

   A. Time

   B. Personnel

   C. Working space

   D. Budget

   E. Inventory

   F. Scope

9. A road construction project is going to require the company's road paver, a piece of equipment that lays asphalt on roadways, in the next two weeks. The road paver is in use until this Friday in a different city and will require five days to be relocated. This is an example of which type of constraint?

   A. Environment

   B. Scheduling

   C. Scope

   D. Quality

10. In the determination of the project scope, which of the following constraints need to be factored into the discussion? (Choose two.)

   A. Project manager

   B. Predefined budget

   C. Mandated finish date

   D. Competitive advantage

11. Chase is a project manager and has released all of the team members from the project, closed vendor contracts, and archived project documents. In what life cycle phase is the project?

   A. Discovery/concept preparation

   B. Initiation

   C. Planning

   D. Execution

   E. Closing

12. The project charter is prepared and agreed to in which project life cycle phase?

    A. Discovery/concept preparation

    B. Initiation

    C. Planning

    D. Execution

    E. Closing

13. After the project charter is signed, what meeting is held to introduce the project team and stakeholders as well as outline the goals for the project?

    A. Lessons learned meeting

    B. Project introductory meeting

    C. Kickoff meeting

    D. Team building lunch

14. Work produced in the high-level risk assessment should be documented in which of the following?

    A. Work breakdown structure

    B. Project charter

    C. Risk register

    D. Quality control plan

15. The steering committee originally mandated that cost was the most important factor to the project, keeping the project team size lean. As the project drags on, the steering committee shifts and tells the project manager that schedule is the most important factor to the project. This is an example of which type of influence on a project?

    A. Change request

    B. Constraint reprioritization

    C. Scope creep

    D. Interactions between constraints

16. Resource allocation, including assigned equipment, team members, and money to support a project, occurs in which project phase?

    A. Discovery/concept preparation

    B. Initiation

    C. Planning

    D. Execution

    E. Closing

17. A project manager is having problems with one team member who is being insubordinate. The project manager approaches the team member to find out what is going on and determines that a change can be made now that the facts are known. This is an example of which of the following?

A. Forcing

B. Avoiding

C. Confronting

D. Smoothing

18. Developing the project team involves all the following EXCEPT:

A. Developing a team that lasts longer than the project

B. Creating a positive environment for team members

C. Creating an effective, functioning, and coordinated group

D. Increasing the team's competency levels

19. The characteristics of the lower-level WBS include all of the following EXCEPT:

A. WBS components are a further decomposition of project deliverables.

B. WBS components should always happen concurrently with determining major deliverables.

C. WBS components should be tangible and verifiable.

D. WBS components should be organized in terms of project organization.

20. A project manager meets with upset team members to listen to their concerns. After hearing their concerns, the project manager makes some of the team members' recommendations in exchange for the team members accepting other rules. This is an example of which of the following?

A. Forcing

B. Compromising

C. Confronting

D. Smoothing

21. Which of the following is the best definition of a capital expense, or CapEx?

A. Money spent on running the business operations of a company

B. Debt accrued on travel expenses needed to run a business

C. Money spent on long-term physical or fixed assets used in a business's operations

D. Debt accrued through the issuing of long-term bonds

22. Expenditures not expensed directly on a company's income statement and considered an investment are known as which of the following?

A. CapEx

B. OpEx

C. TOR

D. SOW

23. Which of the following is the best definition of an operational expense, or OpEx?

    A. Money spent on running the business operations of a company

    B. Debt accrued on travel expenses needed to run a business

    C. Money spent on long-term physical or fixed assets used in a business's operations

    D. Debt accrued through the issuing of long-term bonds

24. DewDrops has received an invoice from their business partner that asks for the monthly payment for staff time plus travel expenses. Which of the following is the best choice on how to categorize this purchase?

    A. CapEx

    B. OpEx

    C. TOR

    D. SOW

25. Wigit Construction is wanting to expand their headquarters campus by building a new state-of-the-art facility for its design crew. Which of the following would be the best categorization of this spending?

    A. TOR

    B. OpEx

    C. ROI

    D. CapEx

26. What does TOR stand for?

    A. Time of reflection

    B. Terms of reconciliation

    C. Terms of reference

    D. Time of reference

27. Derived from the project scope and charter, what would be included in the terms of reference? (Choose three).

    A. General information

    B. Lessons learned

    C. Objectives/deliverables

    D. Project sign-off

    E. Standards

    F. Ishikawa diagrams

28. What is a governance gate?

    A. A checkpoint between project phases where approval is obtained to move forward

    B. A checkpoint where quality is checked against a previously established criterion

    C. Checkpoints at the beginning and end of the project only

    D. After a project governor is appointed, unplanned interruptions from this project sponsor

29. A seller that has passed through a first phase of onboarding through evaluation using a standard criterion to determine a seller's ability to supply a good or service is known as what?

A. Bidder's conference

B. Prequalified vendor

C. Terms of reference

D. Statement of work

30. Jazzmyn just created a business to provide service desk support augmentation services. She is interested in her company having greater visibility with larger businesses and government agencies. The goal is to position her company so that they have a better chance of earning work. In her research, she should inquire to see if these companies have which of the following offerings?

A. Prequalified vendor lists

B. Terms of reference

C. Conflict resolution parameters

D. Scrum retrospective

31. The calculation of the rate of return for a given investment for a given time period is known as which of the following?

A. Payback period

B. Return on investment

C. Time value of money

D. Earned value

32. WigitCom is deliberating whether to go forward with a project to build a new mobile gaming line connected to a popular science fiction franchise. The executive team is concerned about the high cost of this project. What is the best tool the executive team can use to inform the team about the viability of this project?

A. Earned value

B. Cost performance index

C. Present value

D. Return on investment

33. What is the formula used to compute a return on investment (ROI) for a project?

A. Earned value – planned value

B. [(Financial value – project cost) / project cost] × 100

C. Earned value / actual cost

D. Earned value – project cost * 100

34. An IT company is looking to build a new campus to house development staff as the company continues to grow. The results of the ROI analysis were 35%; should the project proceed?

   A. No, the project will yield a negative return on investment.

   B. Yes, the project will yield a positive return on investment.

   C. No, the project will yield a positive return on investment.

   D. Yes, the project will yield a negative return on investment.

35. An IT company is looking to build a new campus to house development staff as the company continues to grow. The results of the ROI analysis were –6%; should the project proceed?

   A. No, the project will yield a negative return on investment.

   B. Yes, the project will yield a positive return on investment.

   C. No, the project will yield a positive return on investment.

   D. Yes, the project will yield a negative return on investment.

36. As a project manager, which plan would you establish to guide the creation of project files, how project documents are filed, where to archive project records, and when to destroy these documents?

   A. Project plan

   B. Records management plan

   C. Project charter

   D. Data capacity plan

37. Which of the following are the reasons proper management of records is important? (Choose three).

   A. Story estimation

   B. Saving time and effort

   C. Ease of finding information

   D. Protect data from unauthorized access

   E. Branding

   F. Budget reconciliation

38. An animation studio has decided to use the principle of least privilege with regard to letting people see artifacts on the project. The principle of least privilege only allows access to information or computer systems to those individuals who need that access to complete a task or deliverable. This is an example of which of the following?

   A. Access requirements

   B. Assess stakeholders

   C. Records management

   D. Quality assurance plan

39. WigitCom is building a revolutionary new gaming platform. A functional manager from the administrative division assigned one of his team members to the project and is curious about the project. They ask this team member about details of the project. To conform to the project's access requirements, what should the project team member say to the manager?

A. Share all the project details with the functional manager.

B. Let the manager review all existing project artifacts.

C. Inform the manager they are unable to share details about the project.

D. Invite the manager into the secure project space and let them look around.

40. A team has set guidance that project communication can only take place through the company-issued collaboration platform or through company-issued devices. This is an example of which of the following?

A. Identify and assess stakeholders

B. A responsibility assignment matrix

C. Establish accepted communication channels

D. Establish communication cadence

41. The team is now storing all the completed project documentation in a repository to be available for audits and to serve as a guide for similar future projects. Which project life cycle phase is the project in?

A. Discovery/concept preparation

B. Initiation

C. Planning

D. Execution

E. Closing

42. Which of the following activities are not part of the Closing phase of a project? (Choose two).

A. Develop a transition/release plan

B. Close contracts

C. Validate deliverables

D. Implement organizational change management

E. Release resources

43. What does the acronym RACI stand for?

A. Responsible, accountable, consulted, and informed

B. Responsibility, authority, consult, and inform

C. Responsible, authority, consulted, and inform

D. Responsibility, accountable, consult, and informed

44. In what project artifact would you find information relating to the quality and availability of resources?

   A. Project schedule

   B. Organization chart

   C. Resource calendar

   D. Risk register

45. All of the following are examples of project resources EXCEPT:

   A. Team members

   B. Equipment

   C. WBS

   D. Materials

46. The launch of a new flagship hotel is coming down to the wire. The customer wants to add an additional welcome area to the hotel prior to the grand opening. Which of the following impacts would be true?

   A. The scope of the project has changed but not the schedule, so costs will increase.

   B. The schedule of the project has changed but not the costs, so the scope will change.

   C. The costs of the project have changed but not the scope, so the schedule will change.

   D. This change will not have a project impact due to contingencies.

47. Where would the following information be found: types of contracts the project will use, authority of the project team, and information on how multiple vendors will be managed?

   A. Budget

   B. Procurement plan

   C. WBS

   D. Detailed risks

48. Deliverables are an output of which phase?

   A. Discovery/concept preparation

   B. Initiation

   C. Planning

   D. Execution

   E. Closing

49. When breaking down project deliverables, what is the lowest level that is recorded in a WBS?

   A. Daily work schedules

   B. High-level requirements

   C. Work package

   D. Major milestones

50. Which of the following are conflict resolution techniques? (Choose two.)

   A. Threatening

   B. Smoothing

   C. Storming

   D. Norming

   E. Negotiating

51. In what stage of team development do team members begin to confront each other and vie for position and control?

   A. Forming

   B. Storming

   C. Norming

   D. Performing

   E. Adjourning

52. Where would an organization document the results of their buy versus build analysis?

   A. WBS

   B. Budget

   C. Change management plan

   D. Procurement plan

53. The following deliverables/activities all occur in the Initiation phase EXCEPT:

   A. Project sign-off

   B. Project charter

   C. Business case

   D. Records management plan

54. Which of the following project documents are created during the Planning phase of a project? (Choose three.)

   A. Status reports

   B. Communication plan

   C. Transition/release plan

   D. Lessons learned

   E. Project schedule

   F. Action items

55. What are project requirements?

   A. A measure of the distance traveled on a project

   B. Characteristics of deliverables that must be met

   C. Checkpoints on a project to determine Go/No-Go

   D. Major events in a project used to measure progress

56. What are critical elements that need to be included in the project schedule?

    A. Define activities, sequence activities, estimate resources, estimate duration

    B. Define activities, budget activities, estimate resources, estimate completion

    C. Budget activities, estimate resources, determine milestones, estimate completion

    D. Develop schedule, determine completion date, check stakeholder assumptions, conduct feasibility assessment

57. Which of the following is the sprint planning meeting used to do?

    A. Get a head start on the work needed for the project.

    B. Prepare the project charter and kickoff meeting.

    C. Set a realistic backlog of items completed during this iteration.

    D. Set the communication and quality plans for the project.

58. The preliminary scope statement should be included in which project document?

    A. Communication plan

    B. Project schedule

    C. Project charter

    D. Lessons learned

59. Which of the following is the main activity of the Execution life cycle phase?

    A. Performance measuring and reporting

    B. Creating and verifying deliverables

    C. Key stakeholder identification

    D. Determining needed project resources

60. What does the acronym RASI stand for?

    A. Responsible, authority, superior, and inform

    B. Responsibility, accountable, support, and informed

    C. Responsible, accountable, superior, and inform

    D. Responsibility, authority, support, and inform

61. You would expect the WBS dictionary to contain all of the following information EXCEPT:

    A. Explanations of team members' roles and responsibilities

    B. Description of the work of the component

    C. Quality requirements

    D. Required resources

62. Even though the project scope statement has been approved, the customer has routinely asked for more features to be added to the product, causing the due date and resources to be adjusted consistently. This is an example of which type of influence?

A. Change request

B. Constraint reprioritization

C. Schedule constraint

D. Scope creep

63. A scope management plan contains which of the following elements? (Choose three.)

A. Process for creating the schedule

B. Process for creating the scope statement

C. Definition of how the deliverables will be validated

D. Process for creating, maintaining, and approving the WBS

E. Process for creating the budget

64. In an Agile methodology, what is a user story?

A. Key information about stakeholders and their jobs

B. Short stories about someone using the product or service

C. Customer survey results after product release

D. Visual representation of product burndown

65. When evaluating the project life cycle phases, in which phase will project costs be the highest?

A. Discovery/concept preparation

B. Initiation

C. Planning

D. Execution

E. Closing

66. What part of a project request defines the reason for the project, the deliverables at a high level, and the project objectives?

A. Work breakdown structure

B. High-level risks

C. Business case

D. Preliminary scope statement

67. A high-level or preliminary risk assessment includes all the following EXCEPT:

A. Risk identification and rolled-up (or categorized) work-task groups

B. Cost–benefit analysis weighing relative risks to potential gains

C. High-level responses to mitigate impact

D. Risk identification of all potential project alternatives

68. The project team has completed all the deliverables for the project. They have meetings scheduled to begin the handoff from the project team to the ongoing operations team. In what project life cycle phase is the project at this point?

    A. Discovery/concept preparation

    B. Initiation

    C. Planning

    D. Execution

    E. Closing

69. Asking questions such as "Does a task have multiple dependencies?" or "Does the task utilize new or unfamiliar technology?" is a step in which process?

    A. Quality assurance

    B. Risk identification

    C. Building the project charter

    D. Building the project schedule

70. In what stage of team development do things begin to calm down because the team members become more comfortable with one another?

    A. Forming

    B. Storming

    C. Norming

    D. Performing

    E. Adjourning

71. Which of the following are ways to organize the WBS? (Choose three.)

    A. Critical path

    B. Subprojects

    C. Project phases

    D. Prioritized by risk

    E. Major deliverables

72. A project assumption can best be described as which of the following?

    A. Internal or external factors affecting the project team

    B. Factors that restrict the project

    C. Factors considered to be true for planning purposes

    D. Factors considered to be true for control purposes

73. How does a high-level scope definition help the planning of a project?

    A. It creates a shared understanding of what is included and excluded from the project.

    B. It sets exactly what a product or service will do.

    C. It is so high level that it ensures that multiple changes can be accommodated by the project.

    D. It helps to shift the blame to the project sponsor if the project is unsuccessful.

74. A project manager meets with team members who are upset. They discuss areas where there is agreement with each other and the situation. Work then resumes on the project. This is an example of which of the following?

A. Forcing

B. Avoiding

C. Confronting

D. Smoothing

75. All of the following are created during the Planning phase of a project EXCEPT:

A. Project schedule

B. Communication plan

C. Lessons learned

D. Change management plan

76. Mitch is a project manager working in the Planning phase of the project. After completing a skills matrix to understand what kind of talent the project will need, he is screening the résumés of talent already employed by the company. What activity is Mitch performing at this stage of the project?

A. Team building

B. Team selection

C. Conflict resolution

D. Trust building

77. Wigit Construction has been using a steamroller to repave a road. A different project finishes early with its paving machine and sends it to the first project to complete the work faster. What type of project change does this represent?

A. Resource change

B. Timeline change

C. Requirements change

D. Quality change

78. What is the project artifact that breaks down resources by category and type?

A. Resource breakdown structure

B. Organizational breakdown structure

C. Equipment breakdown structure

D. Work breakdown structure

79. When the schedule slips on a project because the work is taking longer than planned, what type of common project change does this represent?

A. Risk event

B. Requirements change

C. Funding change

D. Timeline change

80. What are major accomplishments of the project or key events known as?

A. Milestones

B. Deliverables

C. Requirements

D. Risks

81. WigitCom has a key programmer who must leave the project due to a family emergency. The project manager has been able to find another individual who has the experience and skills to take over, and that person is added to the project. This type of communication trigger is known as which one of the following?

A. Project planning

B. Resource changes

C. Incident response

D. Stakeholder changes

82. Among the following, who would be the target audience that should be notified that a milestone has been completed? (Choose three.)

A. Project sponsor

B. Auditors

C. Project team

D. Product end users

E. Steering committee

F. Shareholders

83. In the creation of a work breakdown structure, the project managers ask to see the OBS as an input to creating the document. What is an OBS?

A. Office of business structure

B. Operational business support

C. Organizational breakdown structure

D. Office breakdown structure

84. In which project life cycle phase would you develop a transition or release plan?

A. Discovery/concept preparation

B. Initiation

C. Planning

D. Execution

E. Closing

85. WigitCom has a project underway to modernize its network infrastructure. There are new equipment and standards the network team will need to follow once the cutover is complete. The project team is working to design the knowledge transfer and skills development for the network team once the project is complete. Which activity of the transition/release plan is being developed?

    A. Operations training

    B. Operations handoff

    C. Operations readiness

    D. Skills matrix

86. The team is working on the checklist and activities needed to take a project from development into production. This can include communication, cutover of equipment, promoting code to production, and shutting down the replaced equipment. What part of the transition/release plan is being developed?

    A. Operational training

    B. Go live

    C. Operations handoff

    D. Scrum retrospective

87. DewDrops has been asked to augment a project to develop a new mobile app with outside resources. Once they have selected and onboarded the additional team members, there is a series of training on expectations of the resource management plan, expense, and time reporting, as well as coding standards that will be followed. This is an example of which of the following?

    A. Operational training

    B. Operational handoff

    C. Define access requirements

    D. Train project team members

88. When assessing the resource pool of available talent for a project, what is the purpose of conducting a preliminary procurement needs assessment?

    A. Determine which staff resources can be fired.

    B. Determine options available to augment project staff with needed skills.

    C. Determine the work backlog for the project.

    D. Determine the solution design by selecting a vendor.

89. The Wigit Construction project manager is meeting with new owners of a residential home being constructed. The new owners identify the house must be east facing, have four bedrooms, three bathrooms, a two-car garage, and a fenced-in backyard for the family's dog. What element of the project charter is being developed here?

    A. Establish the ROI of building a house

    B. Project objectives

    C. Success criteria

    D. Preliminary scope statement

90. A project is going to develop new intellectual property for a company that will give it a competitive advantage in the market. The team in considering moving the project work to a secure location that will only allow entry to those with an authorized entry badge. This is an example of which of the following?

    A.   Responsibility assignment matrix

    B.   Baselines and milestones

    C.   Solution design

    D.   Access requirements

91. As a new project is being stood up, the stakeholders and project manager review a feasibility assessment that was conducted on the project idea, review the project management library for similar projects the company has attempted, and review the company's budget and staffing plan for the year. Which activity of the Initiation phase is occurring in this scenario?

    A.   Identify stakeholders

    B.   Review existing artifacts

    C.   Perform an initial risk assessment

    D.   Determine solution design

92. WigitCom is developing a complex new enterprise software platform that will need to be able to be installed on multiple different operating systems and use different browsers. They outline the steps needed for testing in each of the unique environments, which staff will perform the testing, what steps the developers will need to take to meet standards, and how they will keep track of software versions. Which Planning life cycle phase activity are these tasks supporting?

    A.   Review existing artifacts

    B.   Identify stakeholders

    C.   Develop a quality assurance plan

    D.   Determine solution design

93. When developing a transition plan/release plan, which of the following are the appropriate target audiences for the plan(s)?

    A.   Project team only

    B.   Internal audience

    C.   External audience

    D.   Both B and C

    E.   Both A and B

94. A company is implementing a new enterprise resource planning tool that will allow great visibility of financial data as long as the workforce moves away from existing work processes and tools. The project team is working on a plan that will introduce training to the appropriate stakeholders, follow up at intervals after the product cutover, communicate issues and lessons learned, and develop knowledge bases for self-service. These are all elements of which part of the Execution phase?

    A. Organizational change management

    B. Project change management

    C. Organizational standards

    D. Gate reviews

95. DewDrops has contracted with I'm Here for You consulting to provide programming services in the development of a new product. Code turned over to the DewDrops team does not follow the programming standards need for product. The project manager rejects the submission and directs the consultants to correct their defects. This is an example of which of the following?

    A. Obtain project sign-off

    B. Conduct gate reviews

    C. Organizational adoption

    D. Enforce vendor rules of engagement

96. Ron in purchasing needs you to create an SOW prior to releasing an RFP. What does SOW stand for?

    A. Service of workforce

    B. Statement of work

    C. Statement of workforce

    D. Support of work

97. Beth has been charged with gathering information from suppliers to have them bid on specific products or services. What is the appropriate procurement vehicle for this activity?

    A. RFI

    B. Sole source

    C. RFP

    D. RFQ

98. What are the three types of estimates used in three-point estimates?

    A. Fastest-Schedule, Least Resources, Most Desirable

    B. Fastest-Schedule, Optimistic, Most Desirable

    C. Most Likely, Least Resources, Fastest Schedule

    D. Most Likely, Optimistic, and Pessimistic

99. A key stakeholder, Don, is very detail-oriented and likes to do deep dives on the status of projects. He has asked for a highly detailed status report to be emailed to him weekly so that he can stay informed about the project. Which of the following specific stakeholder communication requirements is he asking for? (Choose two.)

A. Frequency

B. Level of report detail

C. Confidentially constraints

D. Criticality factors

100. WigitCom is going to experience a delay in a project because a key programmer is going to be out on emergency leave. What type of communication trigger does this represent?

A. Incident response

B. Business continuity response

C. Schedule change

D. Gate review

101. Joyce is an executive who is not a big user of technology, and she has asked to receive all updates verbally rather than through email. This stakeholder's communication requirement is which one of the following?

A. Frequency

B. Level of report detail

C. Confidentiality constraints

D. Types of communication

102. Chuck is the project manager for a project, and he needs a specialized piece of heavy equipment so that he can get work scheduled. Which project artifact should Chuck look to when checking availability?

A. Project charter

B. Resource calendar

C. Project schedule

D. Project calendar

103. The situation where the project team is stuck on the last piece of work, which prevents the project from completing, is known as which one of the following?

A. Pareto diagram

B. The 95 percent phenomenon

C. IRR

D. The 80/20 rule

104. The practice of documenting who updated a document when and with what information is known as which one of the following?

- **A.** Progressive elaboration
- **B.** Validation
- **C.** Documentation
- **D.** Version control

105. What is the entity that reviews change requests, evaluates impacts, and ultimately approves, denies, or defers the request?

- **A.** CAB
- **B.** CCB
- **C.** NDA
- **D.** MOU

106. A contractor needs to ensure that the subcontractor on the project fulfills the expectations of the customer and wants to create SLAs with the subcontractor. What is an SLA?

- **A.** Service-level agreement
- **B.** Statement logistical alignment
- **C.** Support-level agreement
- **D.** Service-level assignment

107. Who is the target audience and what is the rationale behind conducting a gate review?

- **A.** Project manager, to ensure resource availability
- **B.** Project sponsor, to see if they picked a good project
- **C.** Project team, to find out what they are doing wrong
- **D.** Project governance, to ensure accountability and objective alignment

108. DewDrops's project manager has assembled the project team together to review the proposed project schedule and receive feedback on the feasibility of the schedule. What was the trigger for this communication?

- **A.** Risk register updates
- **B.** Project change
- **C.** Project planning
- **D.** Schedule changes

109. During a project to replace a telephone switch, the project team discovers that one piece of equipment is faulty and a new router will need to be configured. The project manager prepares an email to the organization saying the outage will be extended by an hour to deal with the faulty equipment. What was the trigger for this communication?

- **A.** Task initiation
- **B.** Task completion
- **C.** Project change
- **D.** Schedule change

110. A key stakeholder has considered a project and discovered a new risk that was not previously identified. The project manager shares this information with the appropriate stakeholders, including the project team. What was the trigger for this communication?

A. Project change

B. Risk register update

C. Milestone

D. Task completion

111. When a project manager is working on the creation of artifacts like the scope, budget, and schedule of a project, what type of communication trigger does this represent?

A. Audit

B. Project planning

C. Project change

D. Milestones

112. A project has fallen behind schedule, and the project manager has decided to have team members work 10 hours of overtime each week instead of bringing in new team members. This is an example of which one of the following?

A. Crashing

B. Risk avoidance

C. Fast tracking

D. Critical path method

113. Molly is a key stakeholder on a project. She wants a weekly summary every Friday on the project's progress. An urgent issue comes up on Tuesday that could significantly delay the entire project unless a decision is made quickly. What two factors conflict with each other in this scenario?

A. Intraorganizational differences

B. Criticality factors

C. Tailor the method based on the content of the message

D. Personal preferences

114. Devdan received a file from the project manager to review. When he is finished, he saves the document with his initials at the end of the filename and emails it back to the project manager. Adding his initials at the end is a form of which one of the following?

A. Change requests

B. Risk response plan

C. Progressive elaboration

D. Version control

115. The original project plan for the construction of a building called the facility to be 95 percent defect free by the end of the project. The customer wants few hassles after taking ownership of the building, and this changes to 98 percent defect free. What type of project change does this represent?

    A. Scope

    B. Risk

    C. Timeline

    D. Quality

116. The project manager just completed a meeting with the steering committee and announces to the team that the project can move forward. Which of the following initiated the communication?

    A. Gate reviews

    B. Resource changes

    C. Business continuity response

    D. Milestones

117. Ralph is the project manager for the opening of a new ice cream parlor. The owner is always asking for status updates from Ralph and pushes Ralph to finish with a quality store that is on time and on budget. Which role in a RACI matrix would Ralph hold for the entire project?

    A. Responsible

    B. Accountable

    C. Consulted

    D. Informed

118. Mark submits a purchase requisition to his company's procurement section. He must wait for a PO to be cut in order for the vendor to begin work. What is a PO?

    A. Planned objective

    B. Purchase office

    C. Purchase order

    D. Planned order

119. During a weekly project meeting, there is a list kept of any assignments that are handed out during the meeting. What type of project document is this?

    A. Issue log

    B. Action items

    C. Meeting minutes

    D. Communication plan

120. The project's sponsor is reviewing a communication for the project manager. In the document, she reads that the project is on budget, but there was a delay in getting materials from a supplier since the last communication.

Which project document is she reading?

A. Status report

B. Scope statement

C. Risk register

D. Scrum retrospective

121. The project schedule contains all the following elements EXCEPT:

A. Activity start and finish dates

B. Activity duration

C. Activity assignments to resources

D. Authorization for the project to begin

122. Which of the following are a means of communicating the current conditions of a project? (Choose three.)

A. Dashboard information

B. Status report

C. Meeting minutes

D. Meeting agenda

E. Communication plan

F. Project charter

123. When a company is interested in procuring a commodity or service, they conduct a bidding process where suppliers submit business proposals. What is this process known as?

A. RFI

B. Sole source

C. RFP

D. RFQ

124. Susan is working on a project, and she is the lone specialist assigned to complete a deliverable on a task. She will do the work to complete the assignment, and she is the person who must answer for the correct and thorough completion of the deliverable. Which two participation types of a RACI matrix has she been assigned? (Choose two.)

A. Responsible

B. Accountable

C. Consulted

D. Informed

125. At the product launch of a new service, the project team gathers at a local restaurant to celebrate and honor the work they have done. The project sponsor surprises the team members by giving each one a $200 gift card and thanks them for a job well done. The activity of rewarding and celebration typically occurs in which project life cycle phase?

   A. Discovery/concept preparation

   B. Initiation

   C. Planning

   D. Execution

   E. Closing

126. As a project comes to an end, what are two important activities to conduct with project team members?

   A. Monitor performance.

   B. Give performance feedback.

   C. Release team members to their function.

   D. Ensure project adoption.

127. As work activities conclude, what important step should be performed as team members are released from the project?

   A. Remove physical and digital access.

   B. Sell equipment bought for the project.

   C. Invoice customers for the project team's time.

   D. Train project team members.

128. Why is it important to collect feedback from stakeholders at the conclusion of project activities? (Choose two).

   A. It is a formality of the project closeout.

   B. To tell the stakeholders what a great job was done.

   C. To see if the stakeholders' needs were meet by the project.

   D. To find opportunities for improvement for future projects.

129. Why is it important to validate deliverables when a project starts to conclude?

   A. To confirm the project has satisfied the objectives

   B. To ensure the quality of the deliverables is acceptable

   C. To verify timeliness and completeness to pay a vendor

   D. All of the above

   E. A and C only

130. As a project concludes, the project manager meets with the project team to review expenditures for the project and signs off on the project budget. This is known as which of the following?

    A. Return on investment analysis

    B. Updating the project budget

    C. Current state analysis

    D. Budget reconciliation

131. All of the following are information-gathering techniques used in project management EXCEPT:

    A. Brainstorming

    B. Delphi technique

    C. Gantt chart

    D. Root cause analysis

132. The project manager writes a report, one of the final artifacts created, detailing the status of turnover to production, status of all contracts/purchase orders, total spend on the project, and the release of the resources back to their functional jobs. What is this report called?

    A. Budget reconciliation report

    B. Final project charter

    C. Project closeout report

    D. Update scope statement

133. The project sponsor has been away on a business trip overseas. Her first day back, she finds the project manager to learn the status of the project since she has been away. The project manager develops a quick report and stops by to fill her in. What type of reporting is this known as?

    A. Team touchpoints

    B. Ad hoc reporting

    C. Overall progress reporting

    D. External status reporting

134. Robert is a project manager working on an ERP implementation for a local municipality. He sets a weekly team meeting to review progress on assigned deliverables. What type of tracking tool is this known as?

    A. Team touchpoints

    B. Risk reporting

    C. Gap analysis

    D. Ad hoc reporting

135. The project steering committee wants a monthly report that identifies milestones, schedule and budget updates, risks and issues affecting the project, and any personnel issues. What type of report is the steering committee asking for?

   A. Team touchpoints

   B. Overall progress reporting

   C. External status reporting

   D. Gap analysis

136. The DewDrops board of directors asks for quarterly reports from all their projects over a specific dollar amount. The project manager is asked to attend the meeting and give an update on a new datacenter being constructed. What type of reporting does this represent?

   A. External status reporting

   B. Overall progress reporting

   C. Ad hoc reporting

   D. Risk reporting

137. A contract employee has been brought on to help document the network architecture for DewDrops. Bruce notices this employee has missed several key meetings and he is still waiting for a weekly status update on the work he has performed. He reaches out to the company to inform them of these challenges. What is this an example of?

   A. External status reporting

   B. Validation of deliverables

   C. Monitoring performance

   D. Collecting feedback from stakeholders

138. In which project life cycle phase would the brainstorming, evaluation, and impact of risk be assessed?

   A. Discovery/concept preparation

   B. Initiation

   C. Planning

   D. Execution

   E. Closing

139. Nathan is working on a project status report. He is currently tracking where the project is from a schedule and budget perspective and then making a comparison against the original baselines. What type of tracking is Nathan performing?

   A. Team touchpoints

   B. Risk reporting

   C. Overall progress reporting

   D. Gap analysis

140. Gabriel is compiling the impact and probability of known project threats, including identification of active and new threats and opportunities. What type of reporting is he creating?

    A. Risk reporting

    B. Risk planning

    C. Threat identification

    D. Issues reporting

141. The project team believes they have completed all of the activities and deliverables in the Execution phase. They meet with the project steering committee to review progress, deliverables, and the scope of the project. What type of review is taking place?

    A. Overall progress reporting

    B. Risk reporting

    C. Phase gate review

    D. Gap analysis

142. A sprint backlog is composed of which of the following? (Choose three.)

    A. Sprint goal

    B. Gate review

    C. Run charts

    D. Product backlog items

    E. Milestone chart

    F. Action plan

143. All of the following are examples of categories that might be included in a resource breakdown structure EXCEPT:

    A. External

    B. Project management

    C. Organizational

    D. Probability and impact

144. What is a RACI matrix used for?

    A. Determining key performance indicators

    B. Identifying roles and responsibilities

    C. Analyzing risk at a moment in time

    D. Presenting status via a dashboard

145. A subject matter expert who is not directly working on a project receives a phone call to get her opinion on two options facing the team. Which role in a RACI matrix has this stakeholder been assigned?

   A. Responsible

   B. Accountable

   C. Consulted

   D. Informed

146. A government agency has hired a firm to perform work on its sewer system. As a part of the RFP, the agency requires the successful vendor to carry insurance for errors and omissions. What type of risk strategy is this?

   A. Accept, negative risk strategy

   B. Exploit, positive risk strategy

   C. Share, positive risk strategy

   D. Transfer, negative risk strategy

147. Meghan is a project manager. A team member reported to her an inappropriate and unprofessional interaction with a project team member on a different team. Meghan is unsure to whom this person reports. Which project document would help her find the correct person to whom to escalate this issue?

   A. RACI matrix

   B. Request for proposal

   C. Organizational chart

   D. Issue log

148. The communication plan contains all of the following components EXCEPT:

   A. Stakeholder information needs

   B. When information should be distributed

   C. Report on status of deliverables and schedule

   D. How information will be delivered

149. When creating a RACI matrix, how many different positions can/must have the role of accountable?

   A. Zero

   B. One

   C. Two

   D. Three

150. Which project plan sets how information will be shared on the project, what the frequency level is with which it will be shared, and with whom it will be shared?

    A. Meeting agenda

    B. Action items

    C. Project charter

    D. Communication plan

151. Scott is a board director for a nonprofit, and he is looking for an update on the financial aspects of a new project that the organization is attempting. Which document would help bring Scott up to speed on current conditions of the project?

    A. Issue log

    B. Project management plan

    C. SWOT analysis

    D. Status report

152. WigitCom has just completed the prototype for a new application and has received user acceptance of the design. This is an example of which one of the following?

    A. Gate check

    B. Milestone

    C. Deliverable

    D. Lessons learned

153. When creating a RACI matrix, what does it mean to be assigned an "R" for Responsible?

    A. The one who is ultimately answerable for the correct and thorough completion of the assignment

    B. Those whose opinions are needed before work or a decision is undertaken

    C. Those who do the work needed to complete the task or deliverable

    D. Those who are kept up-to-date when work or a decision is completed

154. In which project phase would an issue log generally be developed?

    A. Discovery/concept preparation

    B. Initiation

    C. Planning

    D. Execution

    E. Closing

155. Gerry is a new project manager in a company, and he has been told that this project must go through checkpoints between phases of the project with a steering committee. What is the name for these checkpoints?

    A. Lessons learned

    B. Scrum introspective

    C. Governance gates

    D. Kickoff meetings

156. At the conclusion of a defined work period, Yolanda got the project team together to figure out what went well, what didn't go well, and what improvements could be made. What type of meeting was this?

    A. Scrum introspective

    B. Product backlog

    C. Daily Scrum

    D. Kickoff meeting

157. There is a question about a project decision that was made at a gathering earlier in a project. Where should the project manager look to help recall the decision and the circumstances that were discussed?

    A. Meeting agenda

    B. Project charter

    C. Action items list

    D. Meeting minutes

158. A project stakeholder receives a regular status update for various activities and milestones after they are completed. What role in a RACI matrix has this stakeholder been assigned?

    A. Responsible

    B. Accountable

    C. Consulted

    D. Informed

159. A meeting has just been conducted on a project where the expectations, goals, and objectives of the project have been explained, including the milestones and timelines of the project. Project sign-off is also likely to occur during this meeting. What type of meeting was this?

    A. Lessons learned

    B. Gate check

    C. Status meeting

    D. Kickoff meeting

160. Gayle, the project sponsor, has said that the transit project needs to have four working bus lines operating 18 hours a day when the project is complete. Gayle has given which one of the following?

    A. Criteria for approval

    B. Gate check

    C. Deliverable

    D. Lessons learned

161. Wigit Construction has been asked to erect a building that has two bathrooms, a conference room, and an open space for cubicles. Which of the following do the bathrooms, conference room, and open space all represent?

    A. Deliverables

    B. Constraints

    C. Assumptions

    D. Requirements

162. Lisa is a contract project manager assisting a government agency that is implementing companywide software. Lisa is operating the project with the belief that executive sponsorship will be strong on the project and that all project team members will be free of other responsibilities. This is an example of which one of the following?

    A. Deliverables

    B. Constraints

    C. Assumptions

    D. Requirements

163. A government agency has invited bids on a project that must be completed by the end of year, have a mobile application to interface the information, and be credit card compliant. These are all examples of which one of the following?

    A. Deliverables

    B. Constraints

    C. Assumptions

    D. Requirements

164. In which project phase would an action items list generally be developed?

    A. Discovery/concept preparation

    B. Initiation

    C. Planning

    D. Execution

    E. Closing

165. The project sponsor has asked that the stakeholders be brought up to speed on project deliverables, schedule, risks, and issues. What is the project sponsor asking for?

    A. Meeting agenda

    B. Status report

    C. Action items

    D. Meeting minutes

166. Only two senior engineers are available to work on a project, and one of them is already committed to work on another project until the middle of the next quarter. This is an example of which one of the following?

   A. Deliverables

   B. Constraints

   C. Assumptions

   D. Requirements

167. Assessing the status of the budget, reviewing risks/issues logs, and measuring performance occurs in what project phase?

   A. Discovery/concept preparation

   B. Initiation

   C. Planning

   D. Execution

   E. Closing

168. The project sponsor receives a phone call letting her know that a major task has been completed successfully and that there were no safety issues to report. Which role has the project sponsor been assigned on this task?

   A. Responsible

   B. Accountable

   C. Consulted

   D. Informed

169. Stephanie is a project manager and has worked with the key stakeholders to get agreement on the project charter. At the kickoff meeting, she makes sure that all key stakeholders and sponsors sign the project charter. This is an example of which one of the following?

   A. Formal approval

   B. CYA

   C. Gate check

   D. Lessons learned

170. What are the defining characteristics of a project? (Choose two.)

   A. Has a definitive start and end date

   B. Is assigned to a portfolio

   C. Creates a unique product or service

   D. Is a part of ongoing operational activities

   E. Is part of an organization's strategic plan

171. A project stakeholder has which of the following responsibilities?

    A. Documentation and administrative support, estimation of task duration, soliciting task status from resources, expertise

    B. Vested interest, providing input and requirements, project steering, expertise

    C. Documentation and administrative support, providing input and requirements, project steering, expertise

    D. Vested interest, providing input and requirements, cross-functional coordination, expertise

172. What is a work breakdown structure?

    A. A task-oriented decomposition of a project

    B. A deliverable-oriented decomposition of a project

    C. A graphic representation of tasks and their sequence

    D. A high-level outline of milestones on a project

173. In what organizational structure does a project manager have the most limited authority?

    A. Weak-Matrix

    B. Projectized

    C. Strong-Matrix

    D. Functional

174. During the Closing phase of the project, what two activities are conducted?

    A. Accept project deliverables and perform quality assurance.

    B. Finalize project work and close all vendor contracts.

    C. Manage stakeholder expectations and close all vendor contracts.

    D. Finalize project work and perform quality assurance.

175. All the following are ways to determine whether a project is completed EXCEPT:

    A. When the project manager declares the project is complete

    B. When the project is canceled

    C. When it has been determined that the goals and objectives of the project cannot be accomplished

    D. When the objectives are accomplished and stakeholders are satisfied

176. A start-up company is attempting to compete in an emerging product market. There are constant disruptive technology changes, and the market is shifting in their product tastes. This type of situation would be best served by which of the following?

    A. Agile approach

    B. Projectized environment

    C. Functional environment

    D. Traditional, or waterfall

177. The Closing processes include all the following EXCEPT:

  A. Archiving of project documents

  B. Release of project members

  C. Review of lessons learned

  D. Monitoring of the risks and issues log

178. A government agency has certain bureaucratic steps it must meet before it can move forward with payment to a vendor. What kind of dependency does this represent?

  A. Discretionary

  B. External

  C. Internal

  D. Optional

179. Which of the following project documents are created during the Execution phase? (Choose two.)

  A. Project charter

  B. Communication plan

  C. Issue log

  D. Lessons learned

  E. Action items

180. The communication plan calls for weekly status email updates, monthly status meetings, and semiannual printed newsletters. This plan is laying out which one of the following?

  A. Criticality factors

  B. Types of communication

  C. Level of report detail

  D. Tailor communication style

181. The steering committee sent an announcement to the project team letting them know that only changes to time, money, and scope will require executive review from now on. All other approvals can now be granted by the project manager. What type of organizational change was made?

  A. Project plan

  B. Business process

  C. Risk register

  D. CCB

182. The project sponsor has decided to make a project manager change. Which project-centric document should the incumbent project manager use to get the new project manager up to speed on the project?

    A. Project charter

    B. Action items

    C. Project schedule

    D. Project management plan

183. Monty cannot remember what the key performance indicators are to help measure success on the product the project team is creating. Where can Monty find this information?

    A. Issue log

    B. Scope statement

    C. Communication plan

    D. Action items

184. A project manager meets with an upset team member to listen to their concerns. After hearing their concerns, the project manager maintains business as usual and instructs the team member that they must comply with the rules. The team member agrees to start behaving accordingly. This is an example of which of the following?

    A. Forcing

    B. Compromising

    C. Confronting

    D. Smoothing

185. The list of items that need to be monitored and/or escalated to minimize the impact on the project team is called what?

    A. Issue log

    B. Action items

    C. Risk register

    D. Budget report

186. The Closing phase of a project serves what critical purpose?

    A. Formal acceptance and turnover to ongoing maintenance and support

    B. Performing governance activities and turnover to ongoing maintenance and support

    C. Formal acceptance and producing deliverables

    D. Performing governance activities and producing deliverables

187. What is the mechanism used to communicate on the status of the project budget?

    A. Expenditure tracking

    B. Expenditure reporting

    C. Budget baseline

    D. Work breakdown structure

188. A project manager is having problems with one team member who is being insubordinate. The project manager does not approach the team member and just tries to carry on as business as usual. This is an example of which of the following?

A. Forcing

B. Avoiding

C. Confronting

D. Smoothing

189. In a briefing to the CEO, the project team explains that there is a risk that the company's two biggest competitors might merge. The CEO asks about the likelihood that this will happen. What exactly is she asking for?

A. A probability and impact matrix

B. The risk impact

C. The risk probability

D. The risk register

190. A project manager listens to the concerns of two team members who are upset with each other. After asking questions, listening, and getting them to talk with each other, the project manager gains agreement on a vested interest for all parties and work resumes. The team members agree to start behaving accordingly. This is an example of which of the following?

A. Negotiating

B. Compromising

C. Confronting

D. Avoiding

191. When does an item move from the risk register to the issue log?

A. As soon as the risk as identified

B. When the risk is triggered

C. Never

D. In the creation of the project plan

192. Level 1 of the WBS always represents which of the following?

A. Critical path

B. Prioritized tasks

C. Project

D. Sponsor

193. Acceptance criteria reviews that are used across the project are known as which of the following?

A. Critical to quality

B. Quality gates

C. Kanban boards

D. Deliverables

194. Which role of the project is responsible for working to create the deliverables according to the project schedule?

A. Project stakeholders

B. Project team members

C. Project scheduler

D. Project coordinator

195. Which of the following is defined as anything that restricts or forces the actions of the project team?

A. A requirement

B. An assumption

C. A constraint

D. An influence

196. A project manager is going to conduct a brainstorming exercise to get a list of potential risks on a project. She invites sponsors, core team members, stakeholders, and SMEs. What is a SME?

A. Service material extract

B. Service matter expert

C. Service material expert

D. Subject matter expert

197. The architectural design team has completed the blueprints for a new building and communicates this to the project manager. What type of communication trigger is this?

A. Task completion

B. Task initiation

C. Incident response

D. Resource changes

198. When regular project status meetings begin, what elements of the change process should be regularly reviewed? (Choose two.)

A. Change request log

B. Risk register

C. Issue log

D. Social media

199. Beau is a project manager who schedules a meeting with subject matter experts to work on activity identification and duration estimation. What triggered this communication?

A. Milestones

B. Audit

C. Scrum retrospective

D. Project planning

200. In the development of a project idea, a stakeholder writes down the purpose of the project, what it might cost, and what business value or benefits will be achieved. What is the document being created?

A. Project description

B. Business case

C. Deliverables

D. Project charter

201. DewDrops presents their plan to the CEO where they can decrease defects by 25 percent. The CEO wants to see if they can decrease defects by 75 percent and asks the project team to reanalyze the project. What project variable is the CEO asking them to change?

A. Quality

B. Funding

C. Scope

D. Risk

202. What is a set of quantifiable measures that an organization uses to gauge progress toward project goals?

A. RACI

B. SWOT

C. KPI

D. NDA

203. At the completion of a meeting, the person who was designated as a scribe will type up their notes along with any decisions that were made and any assignments that were handed out. What type of project-centric document is this?

A. Action items

B. Meeting agenda

C. Status report

D. Meeting minutes

204. In which of the following documents would you find information regarding the high-level budget and milestones for the project?

A. Request for information

B. Scope statement

C. Project charter

D. Action items

205. A project manager is having problems with one team member who is being insubordinate. The project manager approaches the team member and spells out how the behavior is inappropriate and how the team member will behave from now on. This is an example of which one of the following?

- **A.** Forcing
- **B.** Avoiding
- **C.** Confronting
- **D.** Smoothing

Chapter
3

Tools and Documentation (Domain 3.0)

1. Which tool would you use to create a visual representation of timelines, start dates, durations, and activity sequence?

   A. Process diagram

   B. Pareto chart

   C. Gantt chart

   D. Histogram

2. Which knowledge management tool would be used for a team to communicate instantly, share information on task ownership and status, and see events for an entire work group?

   A. Wiki pages

   B. Intranet sites

   C. Vendor knowledge bases

   D. Collaboration tools

3. When dealing with a RACI chart, which of the following are true? (Choose three.)

   A. RACI is an acronym for responsible, accountable, consulted, and informed.

   B. RACI is a hierarchical and seniority-based matrix.

   C. RACI is an acronym for reasonable, accountable, conflicted, and informed.

   D. A RACI is matrix-based chart.

   E. A RACI is a form of a SWOT analysis for risk identification.

   F. A RACI is used to identify roles and responsibilities on a project.

4. Once a change request is submitted, where should it be recorded and assigned an identification number for tracking purposes?

   A. Risk register

   B. Issue log

   C. Business process repository

   D. Change request log

5. What type of project management tool is depicted here?

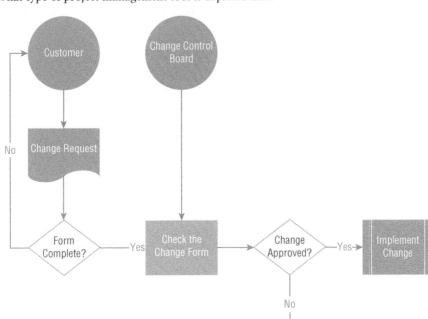

A. Process diagram

B. Histogram

C. Pareto chart

D. Gantt chart

6. What is the comprehensive collection of documents that spells out communication, risk management, project schedule, and scope management?

A. Scope statement

B. Request for proposal

C. Project management plan

D. Dashboard information

7. Amelia is struggling to get a meeting with a key subject matter expert to gather project requirements within the planned timeframe. Where should this information be recorded?

A. Communication plan

B. Issue log

C. Project schedule

D. Action items

8. Shane is the project manager assigned to a project, and he is creating a graphic that shows the project leads for each subject area along with the resources assigned to them. What type of document is Shane creating?

    A. Communication plan

    B. Pareto diagram

    C. Fishbone diagram

    D. Organizational chart

9. Which of the following tools is used for entering data to generate a Gantt chart, WBS, or activity sequence automatically?

    A. Process diagram

    B. Scrum retrospective

    C. Collaboration tools

    D. Project scheduling software

10. Which tool provides a visual representation of all of the steps required in a process?

    A. Process diagram

    B. Histogram

    C. Pareto chart

    D. Project scheduling software

11. Which tool would be used to display observed data in a time sequence?

    A. Run chart

    B. Histogram

    C. Pareto chart

    D. Scatter diagram

12. During a regular check-in meeting with stakeholders, the project manager presents information on all task areas. Each task area has a column for time, cost, resources, risks, issues, and changes, and each item shows a red, yellow, or green indication representing the condition of each task in each area. What project-centric document is being presented?

    A. Process diagram

    B. Gantt chart

    C. Project dashboard

    D. Run chart

13. Earl is a load master for an international shipping company. They have been hired to move construction equipment to a different continent by the end of the month. He is aware that the planes required to carry the largest equipment are currently booked until the end of the month on another assignment. Where should Earl record this information?

    A. Risk register

    B. Action items list

    C. Issue log

    D. Status report

14. A new theater production for Halloween is underway. Josh is the director, Vanessa oversees costuming, Ren is in charge of set construction, and Seth is in charge of marketing and business operations. Where would this information be captured?

    A. Project charter

    B. Communication plan

    C. Project organizational chart

    D. Balanced score card

15. Vanina has taken over as the project manager for a large complex project. She has worked with the project team to get updated status and start dates on tasks. What tool would she use to turn this information into updated reports?

    A. Kanban board

    B. Project scheduling software

    C. Intranet sites

    D. Vendor knowledge bases

16. A team that is geographically dispersed has the need to work on documents together and to have more tools than phone calls by which they can communicate. What is a good option for them to consider using?

    A. Wiki pages

    B. Intranet sites

    C. Vendor knowledge bases

    D. Collaboration tools

17. A long meeting is taking place to document all of the steps that have been taken to build a house in order to try to find improvement opportunities. What type of deliverable will be created?

    A. Gantt chart

    B. Histogram

    C. Process diagram

    D. Pareto chart

18. At a gate check meeting, the committee is presented with a report that shows health in different areas of a project: Schedule = Green, Budget = Yellow, Scope = Green, Risk =Yellow, and Publicity = Red. What type of tool is the committee looking at?

    A. Collaboration tool

    B. Scatter chart

    C. Pareto chart

    D. Balanced scorecard

19. During a project meeting, a key stakeholder looks at a report and sees that the schedule now has a yellow setting when last week it was set at green. What kind of data is the stakeholder viewing?

A. Dashboard information

B. Fishbone diagram

C. Project management plan

D. Risk register

20. What type of project management tool is depicted here?

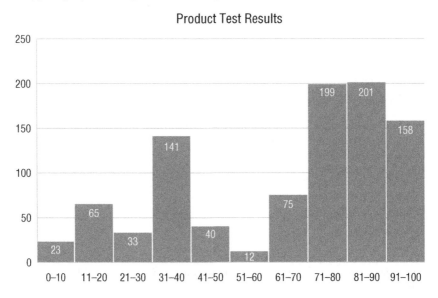

A. Process diagram

B. Histogram

C. Pareto chart

D. Gantt chart

21. Which tool would we use to create a regression line to forecast how the change in an independent variable will change a dependent variable?

A. Process diagram

B. Histogram

C. Pareto chart

D. Scatter chart

22. What kind of tool would automate the creation of critical path, float, WBS, and activity sequence?

   A. Project scheduling software

   B. SIPOC-R

   C. PERT

   D. Run chart

23. Roy is struggling with several issues on a project that are overwhelming productivity. He wants to see which issues are causing the biggest disruption to the project. Which tool should he use?

   A. Histogram

   B. Pareto chart

   C. Run chart

   D. Scatter chart

24. A subject matter expert on a project has been monitoring the distribution of support calls received each hour during the day to determine staffing needs for a project. Which tool is this person most likely using?

   A. Process diagram

   B. Histogram

   C. Run chart

   D. Gantt chart

25. Which type of tool would auto-calculate task duration and planned effort by task, and allow for assignment and reassignment of resources?

   A. PERT

   B. SIPOC-R

   C. Project scheduling software

   D. Histograms

26. A large, global project to create a new movie has over a thousand individuals assigned to the project. The project manager needs to find out which individuals are working with the second production unit. Which project-centric document would help give her this information?

   A. Organizational chart

   B. RACI matrix

   C. Fishbone diagram

   D. Project charter

27. The field team sends an update to the project manager letting him know that there were 12 employees at the project site, there were no accidents this week and no issues to report, and that the team is tracking toward on-time completion of their work. The project manager adds this information in a box on a report. What would you call this data provided to the project manager's report?

   A.   Action items

   B.   Dashboard information

   C.   Organization chart

   D.   Project management plan

28. JoAnn has been tasked with tracking the production data for a prototype product for the first three months in order to ensure that it meets the forecast for quality. Which tool should she use for this analysis?

   A.   Pareto chart

   B.   Run chart

   C.   Fishbone

   D.   Wiki pages

29. Which knowledge management tools allow users to freely create and edit web page content using a web browser?

   A.   Intranet sites

   B.   Wiki knowledge base

   C.   Collaboration tools

   D.   Internet sites

30. Sasha is in a project meeting where a major risk was revealed to be impacting the project. Sasha's boss is in another meeting for the rest of the day, but she needs to provide a detailed update to her boss as quickly as possible. What would be the best method to communicate with Sasha's boss?

   A.   Email

   B.   Distribution of printed media

   C.   Instant messaging

   D.   Text message

31. Wigit Construction is about to begin a project phase replacing a bridge over a major highway. The work will require the highway to be closed at night for several weeks. What is the most appropriate communication method to share these outages with the public?

   A.   Enterprise social media

   B.   Impromptu meeting

   C.   Instant messaging

   D.   Email

32. After establishing the product backlog, what tool would you use to determine the project's velocity?

  A. Burndown chart

  B. Fishbone diagram

  C. Kanban board

  D. Sprint speed

33. The cybersecurity team has been reviewing the security logs and has discovered a potential vulnerability in the software that a project team is developing. The team suggests remediation. What type of communication trigger is this?

  A. Audit

  B. Gate review

  C. Schedule change

  D. Business continuity response

34. What is a graphical representation that shows how much money is left to be spent on a project?

  A. Program Evaluation Review Technique (PERT) chart

  B. Milestone chart

  C. Budget burndown chart

  D. Project network diagram

35. What type of project management tool is depicted here?

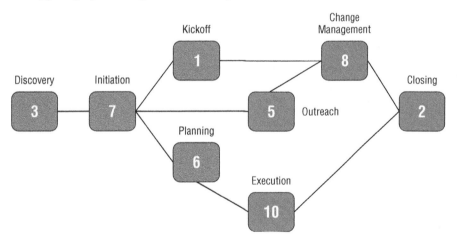

  A. Program Evaluation Review Technique (PERT) chart

  B. Gantt chart

  C. Project dashboard

  D. Budget burndown chart

36. Which tool provides a visual representation of all the steps required in a process?

   A. Process diagram

   B. Histogram

   C. Pareto chart

   D. Project scheduling software

37. What type of project management tool is depicted here?

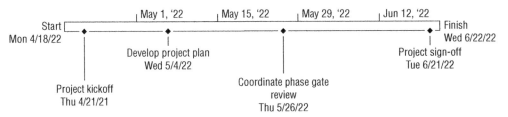

   A. PERT chart

   B. Pareto chart

   C. Milestone chart

   D. Velocity chart

38. What is a collaboration tool allowing online participants to visually create and share ideas?

   A. Enterprise social media

   B. Whiteboard

   C. SMS

   D. Wiki knowledge base

39. What type of project management tool is depicted here?

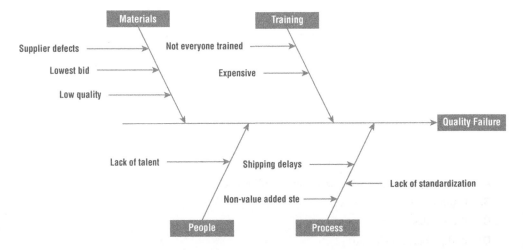

   A. Process diagram

   B. Histogram

   C. Pareto chart

   D. Fishbone diagram

40. The issue log would capture all the following data components EXCEPT:

   A. Issue description

   B. Owner

   C. Mitigation

   D. Due date

41. Which is a special form of a bar chart that visually displays the central tendency, dispersion, and distribution for statistical data?

   A. Kanban board

   B. Histogram

   C. Pareto chart

   D. SIPOC-R

42. Nathan is a software game developer assigned to a project to create a revolutionary new virtual reality game. He receives notice that he is to attend a meeting and give a presentation for 45 minutes on the capabilities of the technology that they will be using. Which document did Nathan receive that communicated this requirement?

   A. Meeting agenda

   B. Meeting minutes

   C. Action items

   D. Wiki pages

43. Mickey is in a project meeting where a risk trigger was identified that will need resolution. Mickey's boss is in another meeting for the rest of the day, but he wants to give a brief update as quickly as possible. What would be the best method to communicate with Mickey's boss?

   A. Enterprise social media

   B. Impromptu meeting

   C. Text messaging

   D. Distribution of printed media

44. The DewDrops project team is conducting a remote meeting to brainstorm ideas for a new application. The team is stuck in their discussion and needing to work freeform to diagram the relationship between different functions. Which is the best tool for the team to use?

   A. Fishbone diagrams

   B. Decision tree

   C. Task board

   D. Whiteboard

45. Which of the following are communication triggers? (Choose three.)

  A. Milestones

  B. Task initiation or completion

  C. Resource changes

  D. Technological factors

  E. Time zones

  F. Interorganizational differences

46. A project team tasked with creating a new software application has discovered that the programming needed is twice as complex as they thought it would be, and they have informed the project manager that they will need more resources or will need to push the due date out by two months. Where would this topic get recorded in project documentation?

  A. Project charter

  B. Issue log

  C. Action items

  D. Risk register

47. What type of project management tool is depicted here?

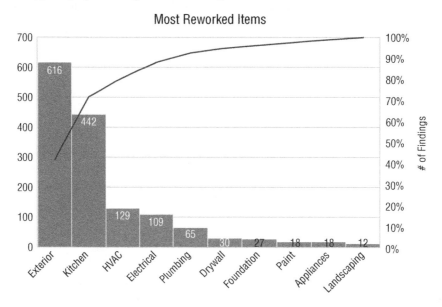

  A. Process diagram

  B. Histogram

  C. Pareto chart

  D. Gantt chart

48. Which of the following is an electronic reporting tool where users can choose elements of the project to monitor project health and status?

A. Kanban board

B. Project scheduling software

C. Vendor knowledge bases

D. Dashboard information

49. Emily has logged into the project website and is looking at various elements of the project health and status. What kind of project-centric documentation is Emily looking at?

A. Meeting minutes

B. Project management plan

C. Action items

D. Dashboard information

50. A project has a team member who routinely is not showing up to meetings or completing assignments. The appropriate place to capture this would be where?

A. Change request log

B. Issue log

C. Risk register

D. WBS

51. What is the only role in a RACI matrix that can only be assigned once per task?

A. Responsible

B. Accountable

C. Consulted

D. Informed

52. A change control template should include all of the following, EXCEPT:

A. The change that is requested

B. The reason for the change

C. The executive sponsor of the change

D. What will happen if the change is not made

53. Which type of chart or diagram is used to represent visually the cause and effect of potential problems to help identify their root causes?

A. Pareto chart

B. Histogram

C. Process diagram

D. Fishbone diagram

54. DewDrops is wondering about the timeliness of their checkout process. Danielle has been tasked with creating a tool to track and display this observed data. What output should she create?

A. Pareto chart

B. Histogram

C. Run chart

D. Scatter diagram

55. Andy is doing a review of a list that contains active problems on the project, their status, owner, and a due date to get resolution. Which project-centric document is Andy looking at?

A. Issue log

B. Organizational chart

C. Action items

D. Status report

56. Interpersonal conflicts are starting to develop on a project. What is the best communication method to use when confronting the problem?

A. Instant messaging

B. Videoconferencing

C. Face-to-face meetings

D. Email

57. An airline engineering company is forced to cut the project budget after poor financial results in the previous quarter. How would this most likely impact the project?

A. The project is postponed due to lack of finincial resources.

B. The project scope is cut back to operate within the new budget.

C. The project will take longer because the number of resources is cut.

D. The project team goes to the steering committee for more funds.

58. When considering a basic communication model, what are the basic elements that are needed? (Choose three.)

A. Receiver

B. Transmission

C. Inbox

D. Non-verbal communication

E. Sender

F. Message

G. Decoder

59. Which type of cost estimating uses a mathematical model to compute costs?

   A. Top-down estimating

   B. Bottom-up estimating

   C. Parametric estimating

   D. Three-point estimating

60. A team is not going to finish a project by the assigned date, and it has asked that three more members be assigned to the work so that the date can be achieved. This is an example of which one of the following?

   A. Issue

   B. Risk

   C. Change

   D. Milestone

61. When a change is being considered, the change control board wants to know how the changes can be reversed if needed. What is this called?

   A. Requirements change

   B. Validation

   C. Version control

   D. Regression plan

62. What are the three types of estimates used in three-point estimates?

   A. Fastest Schedule, Least Resources, Most Desirable

   B. Most Likely, Optimistic, and Pessimistic

   C. Most Likely, Least Resources, Fastest Schedule

   D. Fastest Schedule, Optimistic, Most Desirable

63. All of the following would be the recipients of the communication in a risk register update EXCEPT:

   A. Project manager

   B. PMO

   C. Project sponsor

   D. Project team

64. Which tool would be used to help find the distribution of issues (or other variable) from highest to lowest as bars on a chart?

   A. Scatter chart

   B. Histogram

   C. Pareto chart

   D. Run chart

65. Analogous estimating is where the cost-estimate is developed by which of the following?

   A. Using a mathematical model to compute costs for the project

   B. Calculating the cost of each activity in the work breakdown structure

   C. Using a similar, past project to develop a high-level estimate

   D. Averaging three different estimates of the project cost

66. Money actually spent for a specific timeframe for complete work is known as which of the following?

   A. Cost variance

   B. Planned value

   C. Actual cost

   D. Earned value

67. Which type of cost estimating is done by assigning a cost estimate to each work package in the project?

   A. Top-down estimating

   B. Bottom-up estimating

   C. Parametric estimating

   D. Three-point estimating

68. Analogous estimating is also referred to as which of the following?

   A. Top-down estimating

   B. Bottom-up estimating

   C. Parametric estimating

   D. Three-point estimating

69. Earned value is an indication of which of the following?

   A. The actual cost of completing work in a specific timeframe

   B. The cost of work that has been authorized and budgeted

   C. The value of the work completed to date compared to the budgeted amount

   D. The total sum of sales earned at project completion

70. A construction company is in the middle of a project to build a guest room on a house. The EV value for the project is $7,000, and the actual cost for the project is $9,500. Select the CV for the project and its meaning.

   A. $2,500 and the project is under-budget

   B. $–2,500 and the project is under-budget

   C. $2,500 and the project is over-budget

   D. $–2,500 and the project is over-budget

71. Which of the following is the measure of the cost efficiency of budgeted resources, expressed as a ratio?

   A. AC

   B. EV

   C. CPI

   D. SPI

72. What is the indication of how fast a project is spending its budget?

   A. Fast-tracking

   B. Expenditure tracking

   C. Crashing

   D. Burn rate

73. A project manager for a distributed project team has been gathering a batch of routine organizational messages, project updates, and status reports. What would be the best communication method with which to share this information?

   A. Videoconference

   B. Impromptu meeting

   C. Email

   D. Scheduled meeting

74. The published schedule for a project has a filename of `ProjectSchedule_v1`. The project manager makes an update to the file, and it now is called `ProjectSchedule_v2`. This is an example of which one of the following?

   A. Version control

   B. Communicating changes

   C. Change request logs

   D. Risk register

75. Andy is a project manager for a collocated team, and he just received a mandate from Human Resources (HR) about a lack of compliance from the project team. HR informs Andy that the compliance is important to meet federal law. What type of communication method should Andy use to get this information out to the team?

   A. Videoconference

   B. Impromptu meeting

   C. Email

   D. Scheduled meeting

76. Which of the following is the form in which project schedules are typically displayed?

   A. PERT

   B. Calendar

   C. Gantt chart

   D. Pareto chart

77. All of the following are cost-estimating techniques EXCEPT:

   A. Bottom-up estimating

   B. Program Evaluation and Review Technique (PERT)

   C. Parametric estimating

   D. Analogous estimating

78. Carl is the president of a binding company that won the job to bind the books for a new young adult novel. The publishing company needs 100,000 books delivered in 12 days. At the end of each day, Carl asks for the completed number of books to make sure that it is at 8,333 or higher. What does this number represent?

   A. MOU

   B. AC

   C. SPI

   D. KPI

79. A systematic and independent examination of project procedures, documentation, spending, statutory compliance, and reporting is known as which one of the following?

   A. Project change

   B. Scrum retrospective

   C. Gate review

   D. Audit

80. At the completion of a project sprint, the project team meets to examine what went well, what didn't go well, and what improvements could be made. This is an example of which of the following?

   A. Governance gates

   B. Product backlog

   C. Daily Scrum

   D. Scrum retrospective

81. A project team has scheduled a meeting to discuss a critical milestone that is in danger of being missed. Not all the meeting participants will be near a computer or good network bandwidth for the call. Which of the following is the best tool to conduct the meeting?

   A. Telephone

   B. Video

   C. Face-to-face

82. Which tool should a project team use to identify in what areas people are spending their efforts and how long it is taking to complete tasks?

   A. Kanban board

   B. Time-tracking tools

   C. Version control tools

   D. Requirements traceability matrix

83. A project team is looking for a tool that allows a visual representation of work and its path toward completion. They want to be able to see upcoming tasks, in-progress tasks, and tasks that are completed. Which tool should they use?

   A. Version control tools

   B. Time-tracking tools

   C. Action items list

   D. Task board

84. What type of project management tool is depicted here?

Req ID	Requirement description	Business justification	Status	Comments
1	Add business intelligence to dashboard	Will enable real time visibility to accident data	Done	Dec 28: started Jan 8: Defect reported Jan 30: Defect fixed
2	Create interface to finance data	Will enable managers to see budget and expenditures	In progress	Make sure to adjust for local currency
3	Add mobile view to web page	Will allow team members to view data in the field	Hold	May not be possible with current project budget
4	—	—	—	—

   A. Burndown chart

   B. Run chart

   C. Fishbone/Ishikawa diagram

   D. Requirements traceability matrix

85. Which project tool allows for automated routing of a document for the collection of approvals and sign-offs?

   A. Requirements traceability matrix

   B. File sharing programs

   C. Workflow and e-signature platforms

   D. Multi-authoring editing software

86. Dusty is a project manager and has been asked to create a request for proposal (RFP) for the selection of new monitoring software for a solar panel production company. Which of the following is the best tool for the creation of this document?

   A. Word processing

   B. Spreadsheets

   C. Presentations

   D. Charting/diagramming

87. A project team member needs to record all of their spending so far on the project so they can keep track of where money has been spent and how much budget is remaining. Which office productivity tool is best suited for this function?

    A. Word processing

    B. Spreadsheets

    C. Presentation

    D. Charting/diagramming

88. DewDrops has cut over a new software application and they are receiving complaints about functionality not working correctly. What tool should be used to help keep track of these incidents and the needed follow-up?

    A. Kanban board

    B. Calendaring tools

    C. Ticketing/case management systems

    D. Task board

89. The project team spent a productive meeting working through the process flow for a new module being built for an organization's finance system. They used sticky notes and a whiteboard to map out the process. Which office productivity tool would work best to memorialize the process?

    A. Word processing

    B. Spreadsheets

    C. Presentation

    D. Charting/diagramming

90. Which of the following are used to conduct virtual meetings? (Choose two.)

    A. Face-to-face

    B. Impromptu visits

    C. Videoconferencing

    D. Voice conferencing

91. Jane is the CEO of a company, and there is an update of a security breach at a customer site. She needs to be notified immediately. What is the best method to notify her of the problem?

    A. Social media

    B. Face-to-face conversation

    C. Fax

    D. Text message (SMS)

92. What factors should be considered when scheduling a video or telephone conference?

    A. Determining whether team members are introverted or extroverted.

    B. Recognition of the different time zones/schedules being used.

    C. Make sure writing materials are available in the room.

    D. Ensuring the meeting room has sufficient seating.

93. A project team will have multiple written deliverables requiring review and feedback from subject matter experts, legal, marketing, and enterprise architecture. What tool should be used to help manage all of the feedback?

A. Issue log

B. Version control tools

C. Task board

D. Time-tracking tools

94. A project manager has been asked to speak to the board of directors about the progress and status of their project. It is expected there will be visual slides to help tell the story. Which office productivity tool is likely to be used?

A. Word processing

B. Spreadsheets

C. Presentation

D. Charting/diagramming

95. A project assistant has been asked to get a requirements gathering meeting scheduled with a large group of individuals both internal and external to the company. What type of tools should he use to coordinate everyone's schedule?

A. Real-time surveys/polling

B. Calendaring tools

C. Conferencing platforms

D. Chat

96. A global project team is collaborating on regional requirements related to a new internet product that is being developed. They are in need of a tool that allows them to see each other's work in a central location. Which type of collaboration tool should they use?

A. Calendaring tools

B. Real-time, multi-authoring software

C. Workflow platforms

D. File sharing platforms

97. In what situations would a project select a local installation of project management scheduling tools? (Choose two).

A. When working with a geographically centralized project team

B. When working with a geographically decentralized project team

C. When there is only a single person or small group of individuals on the project

D. When there is a large project team with multiple contributors to schedule information

98. The project adds a consultant to help determine all the steps needed in the invoice payment process for a company. Which tool would the consultant most likely use?

A. Process diagram

B. Pareto chart

C. Project scheduling software

D. Project charter

99. Which of the following tools would you record risks that you identify might impact the project?

    A. RACI chart

    B. Risk register

    C. Risk probability matrix

    D. Issue log

100. What type of project management tool is depicted here?

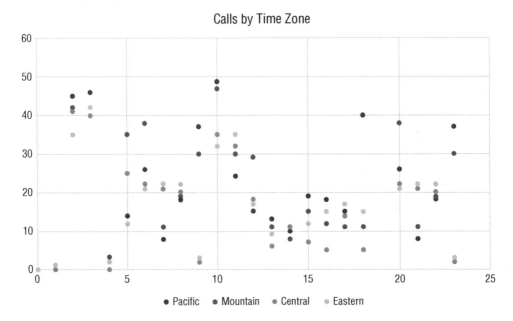

Calls by Time Zone

    A. Process diagram

    B. Histogram

    C. Pareto Chart

    D. Scatter chart

101. What is the list of items that need to be monitored and/or escalated to minimize the impact on the project team?

    A. Scatter chart

    B. Histogram

    C. Pareto chart

    D. Run chart

102. Erik has been doing an analysis to compare any correlation between an independent and dependent variable. He has created a regression line chart to forecast the changes. What has Erik created?

A. Process diagram

B. Histogram

C. Pareto chart

D. Scatter chart

103. A software development team is sharing their knowledge and progress by using a web browser to edit a common page. Which knowledge management tool are they using?

A. Wiki knowledge base

B. Intranet sites

C. Internet sites

D. Collaboration tools

104. A project team has locations in two different cities, and two team members in the different locations are trying to solve a problem. They use software that allows them to share their screens with each other. What type of knowledge management tool are they using?

A. Intranet sites

B. Wiki pages

C. Collaboration tools

D. Internet sites

105. WigitCom is starting a marketing company to build interest in a new mobile device. To help promote the product's "coolness," they have created a series of funny videos about the product. What would be the best method to communicate these videos?

A. Enterprise social media

B. Impromptu meeting

C. Short message service (SMS)

D. Email

106. A road construction company is working on a project to widen 100 miles of road over a six-month period. They are determined to meet the deadline. To do so, they must complete a little over 3.5 miles a week. They start tracking and reporting against this target. What is this an example of?

A. COQ

B. KPP

C. ETC

D. KPI

107. At the completion of a project to implement enterprise software at an organization, the software manufacturer asks the project team to run certain programs that will inspect the number of licenses in use. This is an example of which one of the following?

    A. Project change

    B. Audit

    C. Gate review

    D. Scrum retrospective

108. Rahul has completed his work assignment and is ready to take up a new assignment. He logs on to the team's shared application to update the status of his completed work and to select a new work item. What tool is the team using to help keep track of project work?

    A. Task board

    B. Project scheduling software

    C. Version control tools

    D. Requirements traceability matrix

109. Erin is working on a document that includes an item ID, the type and description of the work that is needed, and an ID for the test cases and their status. What type of tool is Erin creating?

    A. Kanban board

    B. Project dashboard

    C. Requirements traceability matrix

    D. Task board

110. Which office productivity tool would you likely use in the creation of a fishbone or Ishikawa figure?

    A. Kanban board software

    B. Charting/diagramming software

    C. Spreadsheet software

    D. Word processing software

111. A simple spreadsheet that contains an ID number, risk name, risk description, risk owner, and other elements is known as which one of the following?

    A. Risk response plan

    B. Probability and impact matrix

    C. RACI chart

    D. Risk register

112. Jean is looking at a visual representation of the reporting structures that describe the project team member organization. What project-centric document is she looking at?

    A. Project charter

    B. Project schedule

    C. Organizational chart

    D. Communication plan

113. Kim and Bill are co-project managers for a large complex project. They need to calculate and keep track of tasks and due dates, and they must produce reports quickly and accurately.

What should they use for this purpose?

A. Kanban board

B. Daily Scrum

C. Project scheduling software

D. Vendor knowledge bases

114. Which type of document would be given to a committee during a gate check?

A. Status report

B. Communication plan

C. Pareto chart

D. Project scheduling software

115. Scott and his team of network engineers have been working through of list of risks and developing counter actions should those risks appear during the course of a project. They are working on which one of the following?

A. Risk register

B. Risk response plan

C. Probability and impact matrix

D. RACI chart

116. What is the event that detects that a known risk's variable has changed and that it is time to move the item from the risk register to the issue log?

A. Risk response plan

B. Risk identification

C. Risk trigger

D. Probability and impact matrix

117. Jack, Jill, and Harry are three shift managers who help a project work around the clock. What kind of tool would help them keep a log to allow for easy shift change?

A. Enterprise social media

B. Internet sites

C. Wiki pages

D. Vendor knowledge bases

118. A team is not going to finish a project by the assigned date, and it has asked that three more members be assigned to the work so that the date can be achieved. This is an example of which one of the following?

A. Issue

B. Risk

C. Change

D. Milestone

119. A construction project just passed all the electrical inspections for their permits, and the team can now begin to drywall. The project manager shares this information with the work team. What was the trigger for this communication?

A. Task completion

B. Milestone

C. Risk register update

D. Gate reviews

120. Which tool would you use to keep track of the key deliverables and dates important for project continuity?

A. Issue log

B. Risk register

C. Milestone chart

D. Project status report

121. A project is spread across a large city, and a technical problem has developed on the project. There are team members in the field, some located at the headquarters building, and a vendor in a different city. What is the appropriate communication method for this situation?

A. Kickoff meeting

B. In-person meeting

C. Instant messaging

D. Virtual meeting

122. What is a form of mathematical analysis used to shorten the project schedule duration while keeping the project scope the same?

A. PERT analysis

B. Schedule compression

C. Resource smoothing

D. Resource leveling

123. Heather is an audience member at a board meeting scheduled to give details on a project. To help answer questions quickly, she has asked several project team members to stand by to provide information that she doesn't know. What form of communication would they use to share information?

A. Face-to-face meetings

B. Fax

C. Email

D. Instant messaging

124. If a project team is located within the same city, but in different parts of the city, what would be the appropriate communication method to conduct an impromptu meeting?

A. Voice conferencing

B. Instant messaging

C. Daily stand-up meetings

D. Face-to-face meeting

125. What type of project management tool is depicted here?

Task Name	Duration	Start	Finish	4/17	4/24	May 5/1
▲ **Planning**	**11 days?**	**Fri 4/22/22**	**Fri 5/6/22**			
Assess resource pool	3 days	Fri 4/22/22	Tue 4/26/22			
Train project team members	1 day	Wed 4/27/22	Wed 4/27/22			
Develop communication plan	1 day	Wed 4/27/22	Wed 4/27/22			
Develop a detailed scope statement	1 day?	Thu 4/28/22	Thu 4/28/22			
Define units of work	1 day?	Fri 4/29/22	Fri 4/29/22			
Develop project schedule	5 days	Mon 5/2/22	Fri 5/6/22			
Determine budget considerations	1 day?	Mon 5/2/22	Mon 5/2/22			
Develop QA plan	1 day?	Mon 5/2/22	Mon 5/2/22			
Perform initial risk assessment	1 day?	Mon 5/2/22	Mon 5/2/22			
Develop tranistion plan	1 day?	Tue 5/3/22	Tue 5/3/22			
Develop project plan	1 day?	Wed 5/4/22	Wed 5/4/22			5/4
▷ **Execution**	**15.13 days**	**Thu 5/5/22**	**Thu 5/26/22**			

A. Process diagram

B. Gantt chart

C. Pareto chart

D. Scatter chart

126. The EV for a project is 900 and the AC is 1100. The CPI for the project would be which of the following?

A. .82

B. 1.22

C. −200

D. 200

127. MJ has been asked to set up sessions with each of the relevant testing stakeholders so the team can watch the user acceptance testing and gain feedback on areas of improvement. What type of tool should she use to coordinate this event?

    A. Word processing

    B. Print media

    C. Calendaring tools

    D. File sharing platforms

128. What is a list of risks that includes the ID number, name, description, owner, and response plan?

    A. Risk response plan

    B. Issue list

    C. Risk register

    D. Activity log

129. With a major storm approaching, a construction project still has work teams out in the field. The weather service has indicated that everyone should take precautions immediately. What is the best communication method to let the teams in the field know? (Choose two.)

    A. Impromptu meeting

    B. Text messaging

    C. Scheduled meeting

    D. Phone call

130. A status report would contain all of the following information EXCEPT:

    A. Recap of meeting discussion

    B. Tracking deliverables

    C. Condition of project compared to the schedule

    D. Updates on risks and issues

131. During an Agile daily stand-up meeting, what are three questions that are asked and answered?

    A. What did I accomplish yesterday? What will I do today? What are the necessary next steps?

    B. What did I accomplish today? Who will I be working with today? What obstacles are preventing progress?

    C. What did I accomplish yesterday? Who will I be working with today? What obstacles are preventing progress?

    D. What did I accomplish yesterday? What will I do today? What obstacles are preventing progress?

132. What is a document that is distributed prior to a meeting that spells out the topics to be discussed and who will present them at the meeting?

   A. Meeting minutes

   B. Communication plan

   C. Collaboration tools

   D. Meeting agenda

133. What is an inventory of project actions that should be resolved in order to fulfill deliverables?

   A. Histogram

   B. Wiki pages

   C. Action items

   D. Issue log

134. Tony is a project manager, and he gets the team together for a brainstorming activity. At the end of the meeting, they have a list of all the items that need to be completed during the next work period. What type of meeting just occurred?

   A. Scrum introspective

   B. Daily Scrum

   C. Product backlog

   D. Delphi technique

135. Which meeting tool would a project team use for a daily stand-up where some team members are working from home and some are in the office?

   A. Conferencing platforms

   B. Email

   C. Enterprise social media

   D. Wiki knowledge base

136. Debra is having problems getting connected for a conference call through her computer. She calls the Service Desk. They set up an incident to record the break-fix work and to track the remediation of the problem. What productivity tool is the Service Desk using?

   A. Case management system

   B. Conferencing platforms

   C. Real-time surveys

   D. File sharing platform

137. The kickoff meeting for a project needs interactive participation in the modern digital age. They want to allow the participants to use their phones to answer questions and offer their opinions on certain topics. What type of tool would help with this meeting?

   A. Conferencing platforms

   B. Project management scheduling tools

   C. Calendaring tools

   D. Real-time surveys/polling

138. The internal project team, including legal, senior leadership, and finance, are holding a meeting to review their response to a vendor contract. They open the document and are all adding their comments and changes simultaneously. What kind of software platform is assisting this collaborative effort?

   A. Real-time surveys

   B. Whiteboard

   C. Real-time, multi-authoring editing software

   D. Project management scheduling software

139. Greg is working in an oilfield assigned to a project in a remote area. The trailer he works out of has a telephone line, but internet connectivity is limited, and mobile phone coverage is spotty. What would be the best method for Greg to participate in a status meeting?

   A. Email

   B. Voice conferencing

   C. Videoconferencing

   D. Whiteboard

140. What type of project management tool is depicted here?

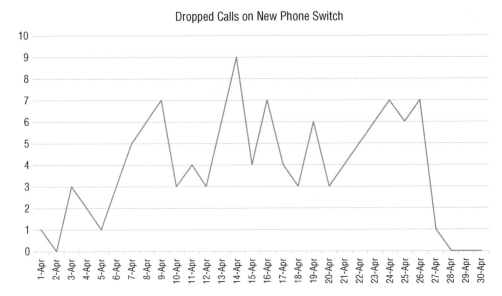

Dropped Calls on New Phone Switch

   A. Run chart

   B. Histogram

   C. Pareto chart

   D. Fishbone diagram

141. A project has an earned value of $2,500 and an actual cost of $2,275. The cost variance for this project would be which of the following?

A. $2,275

B. $225

C. $2,500

D. $–225

142. Which knowledge management tool allows contractors or users of a company's products to find information about a particular issue or to find a work instruction on how to use a product?

A. Collaboration tools

B. Social media

C. Vendor knowledge bases

D. Intranet sites

143. The project management plan consists of all of the following components EXCEPT: (Choose two.)

A. Project schedule

B. Scope plan

C. Action items

D. Communication plan

E. Request for proposal

144. Wigit Construction has reviewed an RFP response and has reached an agreement with the seller. How should they communicate the quantity of goods and services they need and the price that will be paid?

A. Time and materials

B. Purchase order

C. Cost-reimbursable contract

D. Fixed-price contract

145. Hugh has been asked to investigate a new line of medical equipment for a hospital that an executive saw at a trade show. The hospital has no idea who supplies this equipment or how much it might cost. What is the appropriate procurement method Hugh should pursue?

A. SOW

B. RFI

C. RFP

D. RFQ

146. What is a SOW?

    A. A procurement method intended to obtain more information about goods and services

    B. A procurement method designed to invite bids, review, select, and purchase goods or services

    C. A procurement document that details the goods and services to be procured from outside the organization

    D. A meeting with prospective vendors prior to completing a proposal

147. WigitCom is looking for a new storage solution to handle its increasing data needs. They are unclear exactly what kind of solution they might need, or even what vendors could sell them a solution. What is the best procurement method they should use?

    A. Request for proposal (RFP)

    B. Request for quotation (RFQ)

    C. Request for information (RFI)

    D. Statement of work (SOW)

148. DewDrops is looking to solicit bids for the creation of a new manufacturing facility and delivery dock. Which vendor solicitation method should DewDrops use?

    A. Statement of work (SOW)

    B. Request for quotation (RFQ)

    C. Request for proposal (RFP)

    D. Request for information (RFI)

149. The process of submitting a SOW, receiving bids from vendors and suppliers, evaluating responses, and selecting is known as which one of the following?

    A. C&D

    B. RFP

    C. Scrum introspective

    D. RFQ

150. A project team member is spending time typing task information into a computer application that includes tasks, start and end date, and duration. Which project management tool is this team member using?

    A. Collaboration tools

    B. Key performance indicators

    C. Project scheduling software

    D. Project management plan

151. What type of project management tool is depicted here?

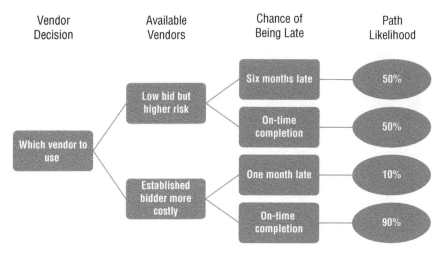

| Vendor Decision | Available Vendors | Chance of Being Late | Path Likelihood |

- **A.** Pareto chart
- **B.** Process diagram
- **C.** Scatter diagram
- **D.** Decision tree

152. A construction company is working on a new building. The CPI for the project is 1.25, which means:

- **A.** The project is over budget.
- **B.** The project is behind schedule.
- **C.** The project is under budget.
- **D.** The project is ahead of schedule.

153. What is the measure of how well a company or project is doing at achieving key business objectives as gauged through a specific value?

- **A.** COQ
- **B.** KPP
- **C.** ETC
- **D.** KPI

154. After cost overruns begin to occur on a project to build a new hospital, a firm is brought in to examine the original budget, expenditures, and the contracts on the project. This is known as which one of the following?

- **A.** An audit
- **B.** A gate review
- **C.** A Scrum retrospective
- **D.** A project change

155. The project manager is working with the project team on a weekly document explaining to stakeholders what tasks have been completed, how much budget has been spent, what issues exist, and any risks that are affecting the project. What document are they working on?

    A. Issue log

    B. Action items

    C. Status report

    D. Project management plan

156. Selina has completed negotiation of a contract with a vendor to create a new mobile application to provide digital identification services for several state governments. What collaboration tool would she use to obtain final approval to complete the contract?

    A. Whiteboard

    B. e-signature platform

    C. File sharing platform

    D. Conferencing platforms

157. Alexander is attending a meeting and the facilitator asks the group to get out their phones and use a QR code to get to a series of questions on their hopes for the meeting. What kind of tool is the facilitator using for this meeting?

    A. Ticketing system

    B. Real-time surveys/polling

    C. Enterprise social media

    D. Real-time, multi-authoring editing software

158. A project team needs to record their user test observations from multiple locations at the same time. They open a spreadsheet application and set up the sheet with the needed rows and columns to capture their results. What kind of platform would allow them all to edit at the same time?

    A. Multi-authoring editing software

    B. Real-time surveys

    C. Conferencing platforms

    D. Ishikawa diagrams

159. Jeff has completed the creation of a statement of work and submits it to his company's procurement section to solicit bids from vendors. The procurement team lets him know that he will need to evaluate responses from vendors and help make a selection.

Which procurement method is described in this scenario?

    A. SOW

    B. RFI

    C. RFP

    D. RFQ

160. Which vendor solicitation method is best to use when more data is required about the goods and services that need to be procured?

   A. Request for information (RFI)

   B. Request for quotation (RFQ)

   C. Request for proposal (RFP)

   D. Statement of work (SOW)

161. Donna is working on procuring services for a project, but she is uncertain about what is available and what the capabilities of various suppliers are to meet this demand. What is the procurement vehicle that she should use?

   A. RFI

   B. Sole source

   C. RFP

   D. RFQ

162. Which vendor solicitation method would be used to invite suppliers into a bidding process to bid on specific products or services?

   A. Request for information (RFI)

   B. Request for quotation (RFQ)

   C. Request for proposal (RFP)

   D. Statement of work (SOW)

163. All of the following are procurement methods EXCEPT:

   A. RFQ

   B. RACI

   C. RFI

   D. RFP

164. DewDrops is the parent company for SunRays, Inc. and DaisyChains, Inc. SunRays and DaisyChains are agreeing to provide service to each other and outline specific criteria each must follow. What is the best method to memorialize the arrangement?

   A. SLA

   B. MOU

   C. NDA

   D. RFP

165. What vendor-centric document would be used to ensure that sensitive or trade secret information is not shared outside of the organization?

   A. MOU

   B. SLA

   C. NDA

   D. RFP

166. When an organization is ready to procure products or services so that work can begin, which procurement method should be used?

    A. Request for information (RFI)

    B. Request for quotation (RFQ)

    C. Request for proposal (RFP)

    D. Statement of work (SOW)

167. What is a bidder's conference?

    A. A daily meeting to ask three questions on project progress and hurdles

    B. A meeting with all prospective vendors to answer questions and clarify issues with an RFP

    C. An auction to help find the lowest cost bid for a product or service

    D. A document that spells out the good or service an organization is looking to procure from outside of an organization

168. DewDrops has issued an RFP, but their statement of work is undeveloped because they are not sure what they want the finished product to do. Which type of contract would make sense in this situation?

    A. Cost-reimbursable contract

    B. Purchase order

    C. Fixed-price contract

    D. Time and materials

169. Wigit Construction is responding to an RFP where the work needed is well-defined, clear, and concise. What is the best type of contract to use if they win the work?

    A. Fixed-price contract

    B. Service-level agreement

    C. Cost-reimbursable contract

    D. Time and material

170. A government agency is looking for a new constituent management system. Though they are unclear about what options are available, they need to be able to make a decision quickly and move forward to procure a solution. What is the best procurement method for them to use?

    A. SOW

    B. RFI

    C. RFP

    D. RFQ

171. DewDrops has issued an RFP, but their statement of work is undeveloped. They decide to use a cost-reimbursable contract with the seller. What best describes the risk condition to DewDrops in this scenario?

  A. Risk doesn't occur until the project is started and doesn't matter at the contract phase.

  B. Risk is evenly divided between buyer and seller.

  C. Riskiest because the buyer won't know the final cost until the project is completed.

  D. Least risky because the seller assumes the risk of unknown costs.

172. Wigit Construction is responding to an RFP where the customer wants to use a fixed-price contract. Which of the following statements represents the risk to Wigit Construction should they win the work?

  A. Least risky because the buyer assumes the risk of unknown costs.

  B. Riskiest because problems on the project will increase their costs.

  C. Risk is evenly divided between buyer and seller.

  D. Risk doesn't occur until the project is started and doesn't matter at the contract phase.

173. The documentation strategy for a project calls for each change to be reflected in a log at the beginning of each document. This is known as which one of the following?

  A. Change request logs

  B. Scrum retrospective

  C. Version control

  D. Risk register

174. An anticipated event of water damage in a construction project is confirmed. The project manager notifies appropriate stakeholders that they will need to bring in a team to perform mold mitigation. What type of communication trigger initiated this communication?

  A. Risk register updates

  B. Business continuity response

  C. Schedule change

  D. Resource change

175. Valentine has been asked to determine the root cause of communication problems on a global project. Which tool should she use to help her with this analysis?

  A. Pareto chart

  B. Run chart

  C. Wiki pages

  D. Fishbone

176. Wigit Construction needs to complete the replacement of six bridges as part of a road-widening project. They need to complete one bridge every 10 weeks for the project to be finished on time. What does the completion of a bridge every 10 weeks represent?

   A. COQ

   B. KPI

   C. ETC

   D. KPP

177. What type of project management tool is depicted here?

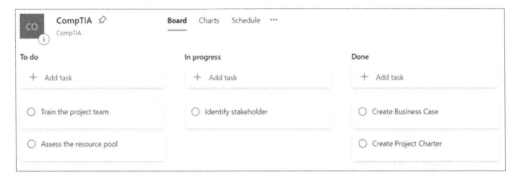

   A. Task board

   B. Requirements traceability matrix

   C. Decision tree

   D. Scatter diagram

178. During a meeting, the team reports that Brad is sorely underperforming. The discussion with Brad reveals he does not have a needed skill set to perform his work. The project manager agrees to pay for an online class so Brad can complete his assignment. Where would this task get captured?

   A. Meeting agenda

   B. Lessons learned

   C. Issue log

   D. Action items

179. Noreen has been working on a document that outlines several pieces of equipment her company needs as well as the professional services required to install the equipment. What document has Noreen created?

   A. Request for information (RFI)

   B. Request for quotation (RFQ)

   C. Request for proposal (RFP)

   D. Statement of work (SOW)

180. A government agency has issued a request for quotation for a new computer solution to help their constituents. What other procurement method is comparable to an RFQ?

   A. RFQ

   B. NDA

   C. RFP

   D. RFI

181. Which two procurement methods serve a similar purpose of scanning the vendor landscape for number of vendors and to gain a cost estimate? (Choose two.)

   A. RFQ

   B. NDA

   C. RFP

   D. RFI

182. What best describes an RFP?

   A. Procurement method to invite bids, review, select, and purchase goods or services

   B. Procurement method to obtain more information about goods and services

   C. A meeting with prospective vendors prior to completing a proposal

   D. Procurement document that details the goods and services to be procured from outside the organization

183. What document would be created to define performance expectations among two or more parties?

   A. MOU

   B. NDA

   C. SLA

   D. RFP

184. Fernando is a project employee whose employment will terminate at the end of the software development project. What vehicle would the company use to protect their intellectual property?

   A. C&D

   B. RFP

   C. KPI

   D. NDA

185. *We are here for you!* is a temporary staffing firm that has sent two administrative assistants to a project to help keep up with demand. What is the best contract vehicle for them to use to deliver this service?

   A. Cost-reimbursable contract

   B. Purchase order

   C. Fixed-price contract

   D. Time and materials

186. DewDrops has issued a document allowing a seller to begin work on a project. The document spells out what the seller will be paid and the exact deliverables that need to be provided on the project. What document has DewDrops issued to the seller?

   A. Purchase order

   B. Cost-reimbursable contract

   C. Fixed-price contract

   D. Time and materials

187. Determining the burn rate and measuring costs to the baseline are elements of what activity?

   A. Expenditure tracking

   B. Spending plan

   C. Parametric estimating

   D. Cost accounting

188. Which type of contract allows the seller to recover all allowable expenses associated with providing the goods or services?

   A. Cost-reimbursable contract

   B. Time and materials

   C. Statement of work

   D. Fixed-price contract

189. The implementation of a new timekeeping system will affect every employee in a company, and they will have to log into a computer each pay period to get paid. What kind of organizational change does this represent?

   A. Outsourcing

   B. Risk response

   C. Internal reorganization

   D. Business process change

190. WigitCom is developing a new product, and it needs a specific skill set to help accomplish the project. They find a staffing firm that can provide them with a resource, but WigitCom is unsure what the specific deliverables of the engagement will be. Which contract vehicle will provide them with the most flexibility?

   A. Time and materials

   B. Purchase order

   C. Cost-reimbursable contract

   D. Fixed-price contract

191. Which type of contract would work best when the product or service needed is well-defined?

    A. Time and material

    B. Scrum methodology

    C. Cost-reimbursable contract

    D. Fixed-price contract

192. Which type of contract would be used in a staff augmentation situation?

    A. Time and materials

    B. Purchase order

    C. Cost-reimbursable contract

    D. Fixed-price contract

193. Which type of contract is riskiest for the seller?

    A. Cost-reimbursable contract

    B. Purchase order

    C. Fixed-price contract

    D. Time and materials

194. Which type of contract offers the greatest flexibility to change scope on a project?

    A. Cost-reimbursable contract

    B. Purchase order

    C. Fixed-price contract

    D. Time and materials

195. Measuring the spending to date, determining the burn rate, and accounting for purchases is known as which one of the following?

    A. Expenditure reporting

    B. Expenditure tracking

    C. Cost accounting

    D. ETC

196. A legal document that describes the goods or services that will be provided, their cost, and any penalties for noncompliance is known as which one of the following?

    A. Request for proposal

    B. Nondisclosure agreement

    C. Contract

    D. Warranty

197. Walt works for Wigit Construction, and he has been assigned as a project manager to build a new bridge. The team has six months to build the project and a fixed budget, and the bridge must handle four lanes of traffic. This is an example of which one of the following?

    A. Three-point estimating

    B. Cost of quality

    C. Triple constraints

    D. Progressive elaboration

198. What best describes an RFI?

    A. Procurement method to obtain more information about goods and services

    B. Procurement method to invite bids, and review, select, and purchase goods or services

    C. Procurement document that details the goods and services to be procured from outside the organization

    D. A meeting with prospective vendors prior to completing a proposal

199. Wigit Construction has completed a project. They submit an invoice to recover the money spent on special marble that they used in the creation of a dining room floor, which the contract allows them to do. What type of contract is Wigit Construction using?

    A. Time and materials

    B. Statement of work

    C. Cost-reimbursable contract

    D. Fixed-price contract

200. The project manager is given a bar chart representing the statistical distribution of traffic on a stretch of road that needs work. What tool has the project manager been handed?

    A. Pareto chart

    B. Histogram

    C. Process diagram

    D. Fishbone diagram

201. What type of project management tool is depicted here?

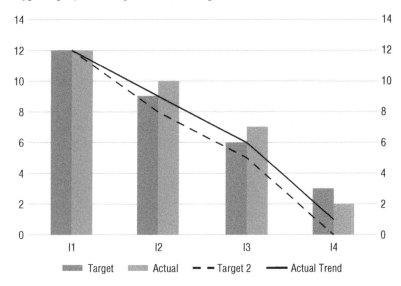

- **A.** Project network diagram
- **B.** Gantt chart
- **C.** Project backlog
- **D.** Project burndown chart

202. A vendor is very strict in applying formal change control on a project for a customer, and few changes are getting approved. Which contract vehicle would cause a vendor to behave this way?

- **A.** Cost-reimbursable
- **B.** Fixed-price
- **C.** Time and materials
- **D.** Request for proposal

203. The project sponsor considers the project to be a failure because the budget continues to climb. What tool can the project manager use to help confirm or deny the project sponsor's suspicion?

- **A.** Balanced scorecard
- **B.** RACI matrix
- **C.** Gantt chart
- **D.** Project management plan

204. Dean has been working on a project for six months, though he is not a permanent employee of the company—he is a contract employee provided by a vendor. Which contract vehicle makes the most sense to use in this case?

    A. Cost-reimbursable

    B. Fixed-price

    C. Time and materials

    D. Request for proposal

205. The project team on which Jenny is working has a detailed set of requirements the project must meet, but the company has limited resources to work on the project. What is the likely result of these two constraints?

    A. The scope must be reduced.

    B. The company must find more resources.

    C. The scope and resources must be adjusted.

    D. The time will expand to deal with scope and cost.

Chapter

4

Basics of IT and Governance (Domain 4.0)

1. Just allowing the consequences of a negative risk to happen is which type of risk response strategy?

   A. Avoid

   B. Transfer

   C. Mitigate

   D. Accept

2. What is the process of examining the risk and establishing the appropriate course of action should it occur called?

   A. Risk analysis

   B. Risk probability

   C. Risk response planning

   D. Risk trigger

3. Kevin is working on a project and he wants to store information on a portable hard drive as a backup to his work computer. What two security considerations should he take into account while using removal media storage? (Choose two)

   A. Back up the data at a regular interval.

   B. Encrypt the removable hard drive to prevent unauthorized access.

   C. Lock the device in an out-of-view area such as a drawer or cabinet.

   D. Do nothing, hackers do not target small projects or businesses.

4. A potential future event that can have either a negative or a positive impact on a project is known as which one of the following?

   A. An issue

   B. A risk

   C. A hope

   D. A requirement

5. As you identify all the potential risks that might impact a project, you should record them in which one of the following?

   A. RACI chart

   B. Risk register

   C. Risk probability matrix

   D. Issue log

6. The activity of selecting risks that have the greatest chance of occurring and the biggest impact on the project should they occur is called which one of the following?

   A. Three-point estimating

   B. Pareto diagraming

   C. Monitoring and Controlling

   D. Risk analysis

7. The consequence or opportunity the risk poses to the project is known as which one of the following?

   A. Risk response plan

   B. Risk impact

   C. Risk register

   D. Risk probability

8. Which of the following is defined as the likelihood that a risk will occur?

   A. Risk response plan

   B. Probability and impact matrix

   C. Risk register

   D. Risk probability

9. Which of the following techniques could be used to create an initial list of risks on a project? (Choose three.)

   A. Parametric estimating

   B. Brainstorming

   C. Three-point estimating

   D. Interviews

   E. Fishbone diagrams

   F. Facilitated workshops

10. What does it mean to share a positive risk?

    A. Assign the risk to a third party who is best able to bring about opportunity.

    B. Monitor the probability or impact of the risk event to ensure that benefits are realized.

    C. Choose to accept the consequences of the risk.

    D. Look for opportunities to take advantage of positive impacts.

11. When determining risk probability and impact, which tool typically offers the best results?

    A. Expert judgment

    B. Parametric estimating

    C. Environmental factors

    D. Project documentation

12. What tool would be used to prioritize and quantify risks, so the information is easy to understand and is visually informative?

    A. Probability and impact matrix

    B. Fishbone diagram

    C. Histogram

    D. Responsibility assignment matrix

13. All of these are common potential risks to a project EXCEPT:

    A. Teams not attending status meetings

    B. Insufficient budget assigned to the project

    C. Scope changes after the project execution begins

    D. Legal ramifications resulting from the project

14. What does it mean to exploit a positive risk?

    A. Assign the risk to a third party who is best able to bring about opportunity.

    B. Monitor the probability or impact of the risk event to ensure benefits are realized.

    C. Choose to accept the consequences of the risk.

    D. Look for opportunities to take advantage of positive impacts.

15. At a minimum, which of the following types of information would be recorded on a risk register? (Choose three.)

    A. Risk score

    B. Risk trigger scores

    C. Risk document review

    D. Risk owners

    E. Description of risk

    F. Mitigation strategy

16. Attempting to ensure that a risk doesn't happen at all, or eliminating the cause of a negative risk, is what type of risk response strategy?

    A. Transfer

    B. Mitigate

    C. Accept

    D. Avoid

17. If a project team wanted to enhance a positive risk, what are they trying to accomplish?

    A. Assign the risk to a third party who is best able to bring about opportunity.

    B. Monitor the probability or impact of the risk event to ensure benefits are realized.

    C. Choose to accept the consequences of the risk.

    D. Look for opportunities to take advantage of positive impacts.

18. What is the activity of determining and documenting any potential risks that might happen on a project?

    A. Risk planning

    B. Risk mitigation

    C. Risk avoidance

    D. Risk identification

19. What is an individual or organization's comfort level with how likely they are to accept or avoid risk?

    A. Risk register

    B. Risk avoider

    C. Risk taker

    D. Risk tolerance

20. Joey has been assigned a project where the deadline must be met before the start of the youth activities season, and there is a limited budget for the project. This is an example of what type of influence?

    A. Scope creep

    B. Interaction between constraints

    C. Constraint reprioritization

    D. Stakeholders opinions on the project

21. Nermit has been working on a chart that lists all the risks that have been identified on a project, along with a numerical score of the likelihood that the risk has of occurring and the score for how impactful the results of the risk occurring would be. What is Nermit creating?

    A. Risk response plan

    B. Risk register

    C. Probability and impact matrix

    D. RACI chart

22. A construction company has been monitoring tropical storms because of the impact a hurricane that might make landfall would have on the project. A tropical storm has just formed, and forecasters are calling for it to make landfall in a populated part of the country. The company begins to buy extra inventory in lumber and other materials. This is an example of which one of the following?

    A. Risk register

    B. Risk trigger

    C. Risk taker

    D. Risk tolerance

23. In which project phase would a risk response plan be activated?

    A. Discovery/concept preparation

    B. Initiation

    C. Planning

    D. Execution

    E. Closing

24. In a briefing to the CEO, the project team explains that there is a risk that the company's two biggest competitors might merge. The CEO asks about the likelihood that this will happen. What exactly is she asking for?

   A.   A probability and impact matrix

   B.   The risk impact

   C.   The risk probability

   D.   The risk register

25. Wigit Construction has completed the risk assessment and cost estimating activities. For certain risks, they have set aside money to cover the costs resulting from possible adverse effects on the project. What are these funds referred to as?

   A.   Contingency reserve

   B.   Ready reserve

   C.   Resource reserve

   D.   Management reserve

26. What are the two forms of acceptance when considering risk response strategies?

   A.   Passive

   B.   Deliberate

   C.   Unintentional

   D.   Active

27. Management for Wigit Construction is concerned with understanding what the negative or positive impacts potential future events might have on a project. What is the management team asking about?

   A.   Contingency planning

   B.   MOU

   C.   Risk

   D.   Issues

28. Every project faces the following potential risks EXCEPT:

   A.   Quality defects

   B.   Not staying on budget

   C.   Scope creep

   D.   Project not finishing on time

29. Which of the following is a tool or technique used in identifying risks to a project?

   A.   RASI

   B.   SWOT

   C.   RACI

   D.   COQ

30. When a change is being considered, the change control board wants to know how the changes can be reversed if needed. What is this called?

   A. Requirements change

   B. Validation

   C. Version control

   D. Regression plan

31. What is the risk response strategy that attempts to minimize the impact or the probability of a negative risk known as?

   A. Avoid

   B. Transfer

   C. Mitigate

   D. Accept

32. WigitCom and DewDrops have entered into a partnership to complete a project. Wigitcom wants to ensure that their trade secrets are not revealed or used by DewDrops. What type of agreement should the two organizations use?

   A. RFP

   B. NDA

   C. SLA

   D. MOU

33. WigitCom is working on a cutover to a new phone system for an agency with a security component. During the planned outage, a problem is discovered that cannot be solved immediately. What should the project team do?

   A. Continue to implement the change.

   B. Implement the regression plan and reverse the changes.

   C. Evaluate the impact and justification of an extended outage.

   D. Identify and document the change.

34. At the completion of a government project requiring security clearance, what document would a vendor likely be asked to sign?

   A. Service-level agreement

   B. Warranty

   C. Request for proposal

   D. Nondisclosure agreement

35. Which of the following roles should be included in the identification of risk on a project?

   A. Subject matter experts (SMEs)

   B. Core team members

   C. Stakeholders

   D. All of the above

   E. Both B and C

36. WigitCom is working on a revolutionary new technology that will potentially alter the entire industry. At an industry conference, a project team member is asked what the team is working on. The employee doesn't share any project details. What is most likely the cause of the employee refusing to share information?

    A. Criticality factors

    B. Cultural differences

    C. Confidentiality constraints

    D. Intraorganizational differences

37. A government agency has two different business units that have a fleet of trash trucks to pick up garbage. To help save money, the agency decides to move the garbage collection function to a single agency and sell the trash trucks of the other agency. What type of business change is this?

    A. Business process change

    B. Outsourcing

    C. Internal reorganization

    D. Staff turnover

38. Which of the following is not true regarding environmental, social, and governance (ESG) factors?

    A. It is imperative the project manager understands ESG as it relates to the project so that regulations, standards, and guidelines are followed.

    B. Awareness of applicable regulations and standards is an ESG factor.

    C. PII is an ESG factor that relates to the social factor.

    D. Project impact to company brand value is an ESG factor.

39. WigitCom is constructing a new mobile ridesharing app. Users of the service will be required to share their location data to sign up for a ride. Which of the following ESG factors would the company need to consider for this project?

    A. National and local privacy laws and regulations

    B. Choosing waterfall versus an Agile approach

    C. Mitigating risk factors

    D. Critical path activities impacting project planning

40. DewDrops is the parent company for SunRays, Inc. and DaisyChains. SunRays and DaisyChains agree to provide services to each other and outline specific performance expectations acceptable to the parent company. Which ESG factors would impact the SunRays and DaisyChains projects?

    A. Submitting to audits and inspections

    B. DewDrops data security access policy

    C. DewDrops's mission, vision, and values

    D. Risk tolerance for each of the three companies

41. What does PII stand for?

    A. Professional infrastructure information

    B. Personally identifiable information

    C. Professional identifiable information

    D. Personal infrastructure information

42. DewDrops has established a project team in another country. Several stakeholders are irritated because of the difficulty in getting approvals to begin work from the provincial government. This is an example of what kind of ESG factor impacting the project?

    A. Project impact to the local and global environment

    B. Awareness of applicable regulations and standards

    C. Awareness of company vision, mission statements, and values

    D. Project impact to company brand value

43. What does PHI stand for?

    A. Personal health information

    B. Professional health initiative

    C. Protected health information

    D. Personal health initiative

44. A new U.S. state law gives customers the ability to request what data a company is storing about their users and how the company is using this information. A global company is having to adjust their product design to conform with this requirement. Which ESG factor is the cause for the project to change in this manner?

    A. Project impact to the local and global environment

    B. Awareness of applicable regulations and standards

    C. Awareness of company vision, mission statements, and values

    D. Project impact to company brand value

45. Which of the following data sources are considered SPII? (Choose two.)

    A. City

    B. Driver's license number

    C. First name

    D. Social Security number (SSN)

    E. Political party

46. *We Are Here for You!* has done work for DewDrops in the past, and has begun to advertise and market that DewDrops is a current project client without DewDrops's consent. What ESG factor might cause DewDrops to issue a cease and desist letter to *We Are Here for You!* to correct this behavior?

    A. Project impact to the local and global environment

    B. Awareness of applicable regulations and standards

    C. Awareness of company vision, mission statements, and values

    D. Project impact to company brand value

47. A project manager is assembling a team for a medical company that looks for the highest standards in ethical behavior. The manager is excited by a candidate's skills, attitude, and experience, but a background check reveals the candidate misrepresented their education and work experience. Which ESG factor likely influenced the decision to move away from this candidate?

   A. Project impact to the local and global environment

   B. Awareness of applicable regulations and standards

   C. Awareness of company vision, mission statements, and values

   D. Project impact to company brand value

48. Which of the following is not true regarding ESG?

   A. PII is an ESG factor that relates to the social factor.

   B. It's imperative the project manager understands ESG as it relates to the project so that regulations, standards, and guidelines are followed.

   C. Awareness of applicable regulations and standards is an ESG factor.

   D. S in ESG stands for social.

   E. Project impact to company brand value is an ESG factor.

49. DewDrops is growing rapidly. The company's leadership wants to install a system that will allow customers to interact with the company to learn more about the company's products and to obtain support. What kind of a system are they considering?

   A. JAD

   B. ERP

   C. CRM

   D. SAFe

50. The risk response strategy that focuses on shifting the liability for a negative risk to a third party is known as which one of the following?

   A. Avoid

   B. Transfer

   C. Mitigate

   D. Accept

51. Risk planning includes all the following activities EXCEPT:

   A. Measuring the SPI and CPI

   B. Analyzing the potential impacts of each risk

   C. Identifying all potential risks to the project

   D. Creating a response to each risk

52. In which phase of a project would you keep an eye on risks to see if any immediate action should be taken?
    - **A.** Discovery/concept preparation
    - **B.** Initiation
    - **C.** Planning
    - **D.** Execution
    - **E.** Closing

53. All of the following are response strategies to positive risks EXCEPT:
    - **A.** Mitigate
    - **B.** Exploit
    - **C.** Share
    - **D.** Enhance

54. All of the following are strategies to deal with negative risks EXCEPT:
    - **A.** Register
    - **B.** Avoid
    - **C.** Mitigate
    - **D.** Accept

55. With a forecast for a worse than average hurricane season, a construction company is aware that certain material costs could rise if a hurricane makes landfall. They begin daily monitoring of the National Weather Service and start actively tracking any tropical storm as they form so they can quickly act to purchase materials in case a hurricane becomes a legitimate threat. What type of risk response strategy is this?
    - **A.** Mitigate
    - **B.** Transfer
    - **C.** Share
    - **D.** Enhance

56. Katie is a project manager whose last performance review just barely met the core standards of the organization. Which of the following choices would most accurately express Katie's risk tolerance?
    - **A.** Risk avoider
    - **B.** Risk decider
    - **C.** Risk taker
    - **D.** Risk observer

57. Karen is a superstar for a company who has had several stellar performance reviews in a row. Which choice would most accurately express Karen's risk tolerance?

A. Risk avoider

B. Risk observer

C. Risk decider

D. Risk taker

58. The project team has done an in-depth root cause analysis as to why certain risks might happen. The impact of these risks would lead to positive outcomes. What type of strategy is this project employing?

A. Accept, negative risk strategy

B. Exploit, positive risk strategy

C. Enhance, positive risk strategy

D. Transfer, negative risk strategy

59. All of the following would be updates to the risk register following a qualitative risk analysis EXCEPT:

A. Causes of risks

B. Watch list of low-priority risks

C. Risks requiring near-term responses

D. Numerical evaluation of each risk

60. The process of determining what impact identified risks will have on project objectives and the probability that they will occur is called what?

A. Qualitative risk analysis

B. Identify risk

C. Quantitative risk analysis

D. Risk categorization

61. There is a section of the project management plan that contains elements of risk, including the project methodology, roles and responsibilities, stakeholder tolerances, and categories. What is this called?

A. Risk register

B. Risk matrix

C. Risk response plan

D. Risk management plan

62. Which of the following are risk response strategies? (Choose two.)

A. Avoidance

B. Assumptions

C. Acceptance

D. Analysis

E. Actual cost

63. DewDrops is working on building the project team, and it is attempting to conduct interviews with company employees located in a different city. Their company practice is to interview candidates in person, but they attempted to conduct video interviews instead due to global travel restrictions related to the ongoing pandemic. What ESG factors influenced this change in direction?

A. Project impact to the local and global environment

B. Awareness of applicable regulations and standards

C. Awareness of company vision, mission statements, and values

D. Project impact to company brand value

64. A company has a large datacenter with nearly a thousand servers and a host of enterprise applications. To align with the company's mission statement, there is a project to move away from high datacenter costs while keeping operational consistency. Which cloud environment should be pursued?

A. PaaS

B. IaaS

C. XaaS

D. SaaS

65. What is the purpose of a data warehouse?

A. To bring various data sources together for the purpose of analysis and decision-making

B. To separate the data's physical location from other applications

C. To handle extremely large data environments

D. To bring all the company's unstructured data together in a central location

66. WigitCom needs a third party to handle hardware and software tools for the application development team to complete a project. Which cloud model should the project team look to use?

A. SaaS

B. IaaS

C. XaaS

D. PaaS

67. What are some disadvantages of using a platform as a service (PaaS)? (Choose three.)

A. Increased pricing at large scales

B. More effective application development

C. Lack of operational features

D. Reduced control

E. Public, private and hybrid options

F. Greater product visibility

68. All of the following are advantages of using a PaaS EXCEPT:

    A. Reduced complexity

    B. Reduced control

    C. Self-ramping up or down of infrastructure

    D. Easier maintenance and enhancement of applications

69. What is a cloud model that offers essential compute, storage, and networking resources on demand using a pay-as-you-go model?

    A. IaaS

    B. PaaS

    C. XaaS

    D. SaaS

70. Which system would be used to help enforce document retention requirements for an IT project's documentation, including project plans and deliverables?

    A. Data warehouse

    B. Enterprise resource planning (ERP)

    C. Customer relationship management (CRM)

    D. Content management system (CMS)

71. All of the following would be considered drivers of adopting a SaaS application EXCEPT:

    A. Decrease the need to update and maintain traditional client-server applications.

    B. Standardization of the HTTPS protocol providing lightweight security.

    C. Reduced costs of developing new software services.

    D. Organization is forced to remain current as the hosting company updates features.

72. What is a software application that manages back-office activities such as accounting, human resources, supply chain, operations, and procurement called?

    A. Enterprise resource planning (ERP)

    B. Customer relationship management (CRM)

    C. Content management system (CMS)

    D. Database management system (DBMS)

73. Which software platform would a company use to easily track all communications and nurture the experience people have with their products and services?

    A. Enterprise resource planning (ERP)

    B. Customer relationship management (CRM)

    C. Content management system (CMS)

    D. Database management system (DBMS)

74. A transit company is looking to set up an application where riders can get from one location to another using a software application to coordinate all the legs of their journey. Which of the following identifies what type of model this is?

A. IaaS

B. PaaS

C. XaaS

D. SaaS

75. In a multitiered architecture, in which tier does the user interact with the system?

A. Presentation

B. Processing

C. Data

D. Application

76. What does MFA stand for?

A. Multifactor architecture

B. Multifactor authentication

C. Mid-tier form architecture

D. Middle factor authentication

77. Kaysie is applying for a job working on a passenger space flight project for a government agency. To help confirm her employment, what steps might be required for Kaysie to be hired as the project manager? (Choose two).

A. Background screening

B. Branding restrictions

C. Service-level agreement

D. Clearance requirements

78. Desiree has pulled into the parking lot at work and is approaching a door where she must scan her badge to gain entrance into the building. Which security concept does this action represent?

A. Facility access

B. Background screening

C. Multifactor authentication

D. Removable media considerations

79. When determining if a project team can view privileged information, which three elements must the team member possess to be allowed to view the information? (Choose three.)

A. Personally identifiable information (PII)

B. A need to know the information

C. Security clearance

    D. Protected health information (PHI)

    E. Access to the information

    F. Industry specific compliance

80. A project is working on a new industry changing technology. To help protect this information, they mark this information as confidential and lock up all key documents and files when there is no one in the room. Why are these documents marked as confidential?

    A. To project national security

    B. To protect trade secrets

    C. To coverup wrongdoing

    D. To identify removable media considerations

81. DewDrops Medical has made a breakthrough in anxiety treatment with the development of a new medication. What is the best way for the company to protect their intellectual property?

    A. File for a patent.

    B. File for a trademark.

    C. File for a copyright.

    D. File for multifactor authentication.

82. *We Are Here for You!* temp agency has hired a firm to create a catchy new jingle that will help customers remember the name and purpose of the company when the company advertises. When they complete the song, what is the best way for the company to protect their intellectual property?

    A. File for a patent.

    B. File for a trademark.

    C. File for a copyright.

    D. File for multifactor authentication.

83. A project team has created an awesome logo for the new product they are creating that identifies the unique device customers can buy. What is the best way to protect their intellectual property?

    A. File for a patent.

    B. File for a trademark.

    C. File for a copyright.

    D. File for multifactor authentication.

84. A project team is working for a national department of defense in the creation of new technology to assist medical teams on the battlefield. What is the best answer on why this information might be classified?

    A. To protect intellectual property

    B. To protect national security information

    C. To protect trade secrets

    D. To establish background screening requirements

85. A project team working on an industry changing technology has a computer policy preventing USB drives from being accessed from the project team's computers. What type of physical security does this policy represent?

A. Data classification

B. Multifactor authentication

C. Mobile device considerations

D. Removable media considerations

86. What are all the reasons a company may lock down the use of removable media on the corporate network? (Choose two.)

A. To prevent malware and virus

B. To make users work experience harder

C. To prevent loss of protected data

D. To comply with the company's mission, vision, and value

87. Which law established protected health information and the rules governing its release?

A. OSHA

B. HIPPA

C. DoCRA

D. PgMP

88. Which of the following would not be considered personally identifiable information?

A. Name

B. Social Security number

C. Date of birth

D. Census data

89. What is the process used to document, identify, and authorize changes in an IT environment?

A. Organizational change management

B. Executive oversight

C. Change control

D. Promote to production

90. The project team is ready to make a change to a web application already in use by an organization. Which artifact should the project manager refer to in rolling out the changes?

A. Downtime/maintenance window schedules

B. Customer notifications

C. Tiered architecture

D. Data warehouse

91. The project team has been asked to identify what actions will be taken if a given IT change has a problem or causes a nonrepairable outage. What has the project team been asked to create?

   A. Automate tests

   B. Risk assessment

   C. Rollback plan

   D. Validation checks

92. David is the IT change manager supporting a large energy company. In support of an upcoming project change, he has established emails to be sent two weeks in advance, a week in advance and the day of the changes so users can expect an outage and a new layout when the service comes back online. What is David working on?

   A. Rollback plan

   B. Notifications

   C. Validation checks

   D. Maintenance windows

93. Jarred is a virtualization engineer on a project that is rolling out a configuration change to the whole server farm. After the change, he runs a script that checks to make sure all servers are back online and that their applications are running. What part of the change control process did he perform?

   A. Check downtime schedule.

   B. Execute a rollback plan.

   C. Execute validation checks.

   D. Send customer notifications.

94. While on a project, Jen has been asked to make a change to an e-commerce application responsible for 90 percent of the organization's revenue. She starts to analyze how long the change might take, what security vulnerabilities exist, and what downstream processes might be impacted by the change. What part of the change control process is Jen performing?

   A. Risk assessment

   B. Automated testing

   C. Release management

   D. Promote to production

95. The project team is setting up an application that will record all the steps in a timekeeping process. They can then run the steps whenever the team updates the software code. What did the project team set up?

   A. Rollback plan

   B. Automated testing

   C. Manual testing

   D. Risk assessment

96. The project team is doing all their work in a development environment. They move any changes to a testing environment for validation before they promote the code to production. What operational change control process concept does this represent?

A. Automate testing

B. Manual testing

C. Tiered architecture

D. Enterprise architecture

97. Eva has completed the software upgrade related to a project-required change. She calls Wally and asks him to verify the changes by logging in and entering a record to confirm the application is working. What is she asking Wally to do?

A. Automated testing

B. Manual testing

C. Risk assessment

D. Requirements definition

98. The project team has been asked to add some additional functionality to an existing application. A business analyst sits down with the stakeholders to be walked through what the application needs to do, how it needs to look, and what other applications or reports the data needs to be shared with. What part of the change control process is the business analyst performing?

A. Requirements definition

B. Customer notifications

C. Release management

D. Rollback planning

99. Denise is part of a change control meeting where she presents her proposed change, the timing and communication to stakeholders, and her rollback plan. She gets a green light to proceed. What was Denise seeking during the meeting?

A. Nothing, this was purely information.

B. Clarification of the maintenance window.

C. Approval of the change.

D. Validation of the change.

100. One project team member has been placed on a performance improvement plan while assigned to a project. The rest of the project team is upset because they do not see management correcting any of this team member's behaviors. What is causing this misunderstanding?

A. Cultural difference

B. Confidentiality constraints

C. Rapport building

D. Criticality factors

101. Which of the following is an example of linkable data becoming SPII?

    A. When the link connects an individual's name to a corporate website

    B. When an individual is shown to live within a specific zip code

    C. When the computer IP address is shown

    D. When an individual's name is linked to their Social Security number

102. A medical laboratory has a secure door that can only be accessed through a retinal scan and by scanning an authorized badge. What security concept is being enacted in this scenario?

    A. Branding restrictions

    B. Scrum retrospective

    C. Legal and regulatory impacts

    D. Facility access

103. Wigit Construction has focused on implementing sustainable practices into its processes and has been hired to build a new bridge over a rail line and a wildlife open space. The triple constraints would imply that the project manager work with time, budget, and scope as it relates to quality in getting this project done. What two aspects of sustainable practices should also govern the project manager's approach to completing this project? (Choose two.)

    A. Ensure pollution prevention in the construction to preserve the wildlife habitat.

    B. Create a fund to help convince local decision-makers to waive environmental regulations.

    C. Create a task force to ensure property rights for the rail line and the wildlife habitat are maintained.

    D. Maintain speed of completion as the most important factor even if crews must work long hours for weeks on end.

104. Time, budget, and scope are considered the triple constraints to tactical project management. What are the three factors that influence sustainable project management from a strategic context?

    A. Environmentally sound, viral social media, economically viable

    B. Economically viable, on time, on budget

    C. Environmentally sound, socially responsible, economically viable

    D. Economically viable, environmentally sound, return on investment

105. Which authentication method requires the user to provide two or more pieces of verification evidence to log into your account?

    A. Certificate-based authentication

    B. Token-based authentication

    C. Multifactor authentication

    D. Biometric authentication

106. For a sustainable-driven company, which of the following is not an area to be evaluated when planning the procurement activities of a project?

   A. Suppliers' past performance or reputation

   B. Unique local requirements of the supplier or project site

   C. Focus on lowest bidder to control costs

   D. The sustainable goals of the provider

107. Risk management focuses on identifying all the positive and negative events that might impact the success of a project or activity. Which of the following would be considered a nontechnical risk?

   A. Project budget is lacking to complete the project.

   B. Local regulations require 30 percent of the project team to live within the country.

   C. The speed to market of the product will determine the market winner.

   D. The time to complete the project is insufficient for the requirements requested.

108. WigitCom has a project to open a store in a new country in which it has never done business before. This new country has laws called the consumer bill of rights, which outline a company's responsibility to collecting private data as well as a consumer's ability to see and/or restrict the collection of this data. Which ESG factor would influence WigitCom's project activities?

   A. Project impact to the local and global environment

   B. Awareness of appliable regulations and standards

   C. Awareness of company vision, mission statements, and values

   D. Project impact to company brand value

109. In May 2018, Europe enacted the GDPR with widespread implications to privacy and the treatment of data. What does GDPR stand for?

   A. Greater Data Privacy and Remediation

   B. General Data Protection Regulation

   C. Greater Discovery for Protecting Remote Access

   D. General Data Privacy Regulation

110. The Payment Card Industry Data Security Standards (PCI DSS) address the handling and management of payment card data. This act covers all aspects of payment card data handling, including acquiring, transmitting, storing, and processing these data. What is the PCI DSS an example of?

   A. Industry-specific compliance

   B. Personally identifiable information

   C. Regulatory impacts

   D. Country-specific privacy stipulation

111. The Sarbanes–Oxley Act is a 2002 U.S. law and was enacted to protect against fraudulent financial transactions by corporations. Sarbanes–Oxley would be an example of which compliance consideration?

A. Industry-specific compliance

B. Protected health information

C. Country-specific compliance regulation

D. National security information

112. Deploying a new information system into production while failing to safeguard security and privacy would have all of these inherent risks EXCEPT:

A. Disclosure of information

B. Unauthorized access to systems and information

C. Failing a project gate review

D. Breach of legal requirements

113. All of the following are used in an Agile approach to project management EXCEPT:

A. Burndown charts

B. WBS

C. Continuous requirements gathering

D. Sprint planning

114. Which information security concept addresses where and how your data is stored?

A. Digital security

B. Physical security

C. Network and system security

D. Application security

115. Sam has been assigned to work remotely in the United States on a project in South America. Why might Sam have to take additional precautions to store data beyond his company's ordinary data practices?

A. To enhance the brand value to the company brought on by the project

B. To protect national security information since data is flowing internationally

C. To align to local security laws governing where and how to store data

D. To take advantage of a new data warehouse being constructed

116. Which of the following are examples of physical security considerations? (Choose three.)

A. Network uptime and availability

B. Smoke and fire alarms

C. Access control systems

D. User authentication

E. Risk management polices

F. Data backups

117. Which of the following is not an area of application security?

  A. Data backup and availability

  B. Data sharing and role-based access control

  C. Project management software data encryption

  D. User authentication

118. Which of the following is both a serious risk management issue and a compliance/legal matter?

  A. Branding restrictions

  B. Anything as a service

  C. Data security classification

  D. Data privacy

119. The commitment to keep a person's or organization's information private and protect unauthorized disclosure of information from being disclosed is known as which of the following?

  A. Confidentiality

  B. Security clearance

  C. Privacy

  D. Data sensitivity

120. Which project management tool helps determine whether a system, process, or program involving personal information raises privacy risks?

  A. Multitiered architecture

  B. Privacy impact assessment

  C. Service-level agreement

  D. Backlog prioritization

121. Crystal has applied for a job supporting a national space program and has been asked to complete documentation that will allow a thorough review of her trustworthiness to work on the project. Which information security concept is in play for this applicant?

  A. Clearance requirements

  B. Facility access

  C. Background screening

  D. Access on a need-to-know basis

122. What is the purpose of establishing security clearances for project team members?

  A. Democratize data to even the playing field for all.

  B. Create needed overhead to keep people employed.

  C. Allow vetted individuals access to restricted information.

  D. Protect the team members' personally identifiable information.

123. Project team members are required to get a badge that opens electronic locks for access to where a new product is being created. Which type of security access does this represent?

A. Physical security

B. Operational security

C. Digital security

D. Application security

124. A project has developed a new political polling application to help in an upcoming election. The application was originally developed for a single state's use, but now there seems to be a market for the application across the country. Which of the following is not one of the actions the project team should take next?

A. Assess legal and regulatory impacts for a national rollout

B. Identify country- and state-specific privacy regulations

C. Enable multifactor authentication for project team members

D. Conduct a privacy impact assessment

125. WigitCom's project manager is ready for the team to promote new code to production. In talking with the business, they decide the third Sunday between 8 p.m. and 11 p.m. would be the most ideal as business is the slowest during that time. What did the project manager and business just negotiate?

A. Promote to production plan

B. Automated testing

C. Customer notifications

D. Maintenance window

126. Which of the following are reasons to build sustainability into project management practices?

A. Reduces resource turnover

B. Reduces project crisis situations

C. Minimizes project cancellations and interruptions

D. Creates a competitive advantage

E. All of the above

F. None of the above

127. A project is underway to expand service to a new country. This new country has laws called the consumer bill of rights governing the collection of private data as well as a consumer's ability to see and/or restrict the collection of this data. Since the project team is part of a global company, would they have to adhere to these regulations?

A. No, as a global company they need to manage their service to the global need and not the local company.

B. No, the company was not doing business when the law was passed.

C. Yes, to avoid fines or a loss of the ability to do business in country, they must conform to the law.

D. Yes, as a gesture of goodwill, the company should start out following the law and then use their leverage later to avoid compliance.

128. A project manager is working on a new ride share system in the United States. The project manager is aware of the GDPR law that was passed in Europe. Although the project is based in the United States, why might the project manager need to adhere to the GDPR?

    A. They would not; the laws of the United States are the only ones that matter.

    B. For ease of scalability, in case the application is ever moved to Europe.

    C. They would not, since the application would never be used in Europe.

    D. When data on European citizens is stored, companies must abide by the GDPR.

129. Why might a company consider choosing to integrate sustainability concepts into their project management practices?

    A. Organizations want to assume the responsibility for the societal impacts of their products and services.

    B. Organizations want to avoid the risk associated with their products and services.

    C. Organizations want their customers to assume responsibility for the societal impacts of their products and services.

    D. Organizations do not want to integrate sustainability into their projects because it is bad for the bottom line.

130. What is the "triple bottom line" in creating sustainability in project management?

    A. Balance between time, money, and scope as it pertains to quality

    B. Harmony between time, scope, and economic sustainability

    C. Balance between social sustainability, environmental sustainability, and scope as it pertains to quality

    D. Harmony between economic, social, and environment sustainability

131. Where should the identification of relevant environmental and social risks and impacts be recorded?

    A. They are not recorded as part of a project.

    B. They are recorded in the risk register.

    C. They are recorded in the meeting minutes.

    D. They are recorded in the issue log.

132. Which ESG factor shows up as a criterion for deciding whether to take on a project and how to execute that project?

    A. Environmental

    B. Social

    C. Governance

    D. None of the above

133. Which ESG factor considers the relationships among stakeholders who take part in or are influenced by a project?

    A. Environmental

    B. Social

    C. Governance

    D. None of the above

134. Which ESG factor directs the way a project will be led as well as the regulations and standards that may dictate how the project will unfold and what processes must be followed?

    A. Environmental

    B. Social

    C. Governance

    D. None of the above

135. Which of the following are reasons to build sustainability into project management practices?

    A. Increases resource turnover

    B. Maximizes project crisis situations

    C. Ensures project cancellations and interruptions

    D. Creates a competitive advantage

    E. All of the above

    F. None of the above

136. Which ESG factor is most likely to get overlooked in project planning and execution?

    A. Environmental

    B. Social

    C. Governance

    D. All of the above

    E. None of the above

137. Questions such as "Will the customer experience service be disrupted?," "Will the project team be expected to work extended hours?," and "How will a new organizational structure affect diversity?" are related to which ESG factor?

    A. Environmental

    B. Social

    C. Governance

    D. Organizational

138. WigitCom planned a project to decommission a datacenter but failed to consider how to properly dispose of retired servers, monitors, and network gear. The original plan had an activity to select the lowest bidder for the disposal, but a local regulation requires this type of equipment be disposed of through a certified electronics recycler. Which two ESG factors contributed to this project change? (Choose two.)

- **A.** Environmental
- **B.** Social
- **C.** Governance
- **D.** Organizational

139. Which of the following would be potential project impacts regarding governance? (Choose two.)

- **A.** Pollution of the local environment's water supply and landfills
- **B.** Project delays due to failing to consider regulatory requirements on the project schedule
- **C.** Rework due to failing to meet standards related to personal data
- **D.** Project team members not being appropriately compensated for overtime

140. For a sustainable-driven company, which of the following is not an area to be evaluated when planning the procurement activities of a project?

- **A.** Suppliers' past performance or reputation
- **B.** Unique local requirements of the supplier or project site
- **C.** Ability to comply with industry standards such as HIPPA or PCI
- **D.** The ability to adhere to meet time deadlines over all other requirements

141. DewDrops just launched a new vision to be the model for sustainable practices in their industry. A project is nearing completion when this new mission was announced through a press release. What steps, if any, should the project team take to account for this change in vision?

- **A.** Do nothing; stakeholders will understand the project was almost complete when this was announced.
- **B.** Put a warning on their product that the product will likely change in the future because of this announcement.
- **C.** Perform a sustainability impact assessment to see if there are deliverables that need to be changed and update the project plan if so.
- **D.** Move forward "as is" and incorporate the new philosophy into future projects.

142. The CEO of a large corporation has asked to be granted access to all digital files for a new project. Why might the CEO be denied this request despite being the highest-ranking individual in the organization?

- **A.** Does not have clearance.
- **B.** Does not have access.
- **C.** Does not have a need to know.
- **D.** This request should always be granted.

143. A software development project has changed project managers from Ellen to Michelle. Using the principle of least privilege, which level of access should Michelle be given to look at the project data?

A. Michelle should be granted the same access as Ellen.

B. Michelle should only be granted high-level access to data until she has been on the project longer.

C. Michelle does not need access granted to her to perform her duties on the project.

D. Michelle should be granted access to all information in the organization.

144. A project manager has asked for a legal review of a contract before it is awarded. The email the project manager receives from the legal team includes language stating "This correspondence represents privileged and confidential communication between a client and their attorney." What type of data security protocol does this warning provide?

A. Multifactor authentication

B. Access on a need-to-know basis

C. Data classification

D. Background screening

145. Anvay is part of the IT security team and has been asked to grant permissions for a new employee to see the engineering team's file share. What step should Anvay take prior to performing this task?

A. Grant the access immediately.

B. Check the business justification and compare to established security protocols.

C. Do not grant the access.

D. Grant the access after an established waiting period.

146. A project team member is being terminated from a project involved with the development of intellectual property effective immediately. What is the appropriate course of action for the project manager?

A. Brief the terminated employee on brand restrictions.

B. Perform a background screening and check clearance requirements.

C. Immediately revoke the terminated employee's permissions.

D. Wait for guidance from human resources before doing anything.

147. A list of steps undertaken to undo a release and restore a system to its original state is known as what?

A. Risk assessment

B. Anything as a service

C. Continuous integration process

D. Rollback plan

148. Which form of quality assurance for a software release would include team members following a checklist and performing actions such as data entry, validating functionality, and viewing reports?

A. Automated testing

B. Turing testing

C. User acceptance testing

D. Manual testing

149. A system needs to be taken offline for maintenance to be performed. The project manager reviews the RACI chart and finds the dispatch center has to be informed about the outage. What change control step should the project manager take?

A. Perform a risk assessment.

B. Send customer notifications.

C. Perform software validation checks.

D. Release the software.

150. A company has gone through a rapid growth cycle and now has several one-off systems for activities like the company's finances, supply, and human resources. The leaders of the company want to have a system that brings these activities together. What kind of system would meet these needs?

A. CRM

B. LAN

C. ERP

D. XaaS

151. WigitCom has incorporated a sustainable focus into their project to build a new videoconferencing platform. Which of the following is not one of the elements the project manager would need to consider when adopting a more sustainable approach?

A. Work-line balance of the project team to reduce burnout

B. Limiting unnecessary travel of the project team to reduce emissions

C. Creating a compensation model to get the most affordable project team

D. Creating equitable hiring practices to support innovation and social inclusion

152. WigitCom has built a new application to allow customers to share their homes to guests who want to use the home for their vacation. They are excited for the opportunities this new software will provide to the property owners and renters alike. Which laws would make the most sense to design against from the inception of the project?

A. GDPR

B. HIPP-A

C. PMBOK®

D. PRINCE2

153. Trevor is the project manager over a civil engineering project, and one of the risks of the project has the potential for a sinkhole in the area where a road is supposed to be built. The project team reports that a sinkhole did in fact form in this area. Which two items are true about this event? (Choose two.)

   A. The risk has now become an issue.

   B. The project team accepted this risk.

   C. The risk trigger has happened, calling for a response.

   D. This needs to go through the change control process.

154. The project team is trying to update a software application that exists on a single-server environment. This is complicated testing, development, quality assurance, and training. What type of solution could they recommend to help the development of this application?

   A. Automated testing

   B. Requirements definition

   C. Tiered architecture

   D. Continuous deployment process

155. WigitCom has a problem on an e-commerce website that is causing an outage resulting in a million dollars of lost revenue every hour. In which environment should changes be made?

   A. Testing

   B. Production

   C. Quality assurance

   D. Development

156. In IT service management, what is the standard traditionally used for change control methods guiding the entire IT life cycle?

   A. Information Technology Infrastructure Library (ITIL)

   B. Project Management Body of Knowledge (PMBOK®)

   C. Business Analysis Body of Knowledge (BABOK)

   D. Awareness Desire Knowledge Ability Reinforcement (ADKAR)

157. Which type of change management focuses on service transitions such as moving from one vendor to another or getting all your departments on board with a new ERP platform?

   A. Organizational change management

   B. On-premises change control

   C. Cloud change control

   D. Shift change management

158. Which change control method is divided into areas of focus such as business model, governance, and the use of a platform?

   A. Organizational change management

   B. On-premises change control

   C. Cloud change control

   D. Shift change management

159. All of the following are challenges with cloud change management EXCEPT:

    A. Siloed communications

    B. Restricted approval process

    C. Rapid approval process

    D. Legacy business processes lacking agility

160. Which change control approach needs to facilitate faster change processes?

    A. Organizational change management

    B. On-premises change control

    C. Cloud change control

    D. Shift change management

161. Which change control approach typically needs abundant planning for capacity upgrades and other infrastructure changes?

    A. Organizational change management

    B. On-premises change control

    C. Cloud change control

    D. Shift change management

162. Which change management philosophy is focused on the adoption of new business process, tools, and support models?

    A. Organizational change management

    B. On-premises change control

    C. Cloud change control

    D. Shift change management

163. Which process refers to the planning, design, scheduling, testing, and deploying of new applications or upgrades?

    A. Cloud change control

    B. Incident or problem management

    C. On-premises change control

    D. Software release control

164. Which type of management is primarily concerned with how changes flow through preproduction environments?

    A. Problem management

    B. Release management

    C. Change management

    D. Project management

165. Wigit Construction has been tasked with repaving a bridge that is a major thoroughfare for a large metropolitan community. They are allowed to shut down traffic from midnight until five in the morning for two weeks straight. What would this five-hour block be called in operational change control terms?

A. Release management

B. Validation check

C. Work breakdown structure

D. Maintenance window

166. Kayla has been working during an outage to upgrade a production database to its most current version. The upgrade is behind schedule, and she has three hours of work to complete in the next 45 minutes. What change control process should she execute?

A. Perform validation checks.

B. Initiate the rollback plan.

C. Start automated testing.

D. Modify the downtime window.

167. The build and unit testing stages of a software release process where every version committed triggers an automated build and test is known as which of the following?

A. Cloud change control

B. Anything as a service

C. Continuous integration/continuous deployment process

D. Tier architecture/promote to production process

168. What is a development process where programmers commit code to repositories frequently and ideally several times a day known as?

A. Continuous integration

B. Software release management

C. Software as a service

D. Enterprise resource planning

169. The collection of hardware and software that determines how fast computations can be made, when systems are connected, and what storage is available is known as which of the following?

A. Infrastructure as a service

B. Multitiered architecture

C. Electronic document and record management systems

D. Computing services

170. What is the correct definition of storage?

    A. The processing power, memory, networking, and other resources needed for computational success of a program

    B. The process through which digital data is saved and recovered through computer technology

    C. A structured set of data held in a computer with multiple channels for access

    D. A connected set of computers used to exchange data and share resources with each other

171. What is the definition of a database?

    A. The processing power, memory, networking, and other resources needed for computational success of a program

    B. The process through which digital data is saved and recovered through computer technology

    C. A structured set of data held in a computer with multiple channels for access

    D. A connected set of computers used to exchange data and share resources with each other

172. In terms of information technology, what is the best way to describe a network?

    A. The processing power, memory, networking, and other resources needed for computational success of a program

    B. The process through which digital data is saved and recovered through computer technology

    C. A structured set of data held in a computer with multiple channels for access

    D. A connected set of computers used to exchange data and share resources with each other

173. The facilities department has all the blueprints and support plans for the corporate buildings they manage. Which of the following would be the appropriate location to store this information?

    A. Enterprise resource planning (ERP)

    B. Customer relationship management (CRM)

    C. Electronic document and records management system (EDRMS)

    D. Database management system (DBMS)

174. May has completed her work for the week. She logs into a system to record her time and billing for the week. When she completes that task, she moves to another part of the system to record her expenses for reimbursement purposes. What kind of system is May logged into?

    A. Data warehouse

    B. Customer relationship management

    C. Enterprise resource planning

    D. Content management systems

175. Max has been assigned as the project manager for a new IT project. She has been assigned an account code to track and report on all budget and spending items related to the project. Which software application does the organization assign and use this code?

   A. Financial systems

   B. Software as a service

   C. Data warehouse

   D. Project management information system

176. The marketing department of DewDrops has just completed a campaign for a service in a new area. The team wants to share their data and findings with the rest of the company to help other departments and the corporate leadership look for trends and help guide future decision-making. Where should the team upload this information?

   A. Data warehouse

   B. Cloud storage

   C. Print the data and distribute hard copies

   D. To a PaaS provider

177. Which cloud model is a software licensing and delivery model where software is licensed on a subscription basis and is centrally hosted?

   A. IaaS

   B. PaaS

   C. XaaS

   D. SaaS

178. What would you call detailed explanations of how to execute a process, method, task, or program that also outlines precise steps needed to successfully complete the item?

   A. Anything as a service

   B. Data warehouse

   C. Documentation

   D. Computing services

179. Which system would you use to process expense reports, make entries into the general ledger, or process PCard (purchasing card) or POs (purchasing orders)?

   A. Content management systems

   B. Data warehouse

   C. Electronic document and record management systems

   D. Financial systems

180. Will works taking phone calls from the public dealing with a variety of issues, including billing, adding or ending service, or complaints about service quality. What type of system would Will be using to research the caller's history and find meaningful information related to their account?

A. Enterprise resource planning (ERP)

B. Customer relationship management (CRM)

C. Content management system (CMS)

D. Database management system (DBMS)

181. Nancy has been hired on to a project to help manage the social media presence and develop a website communicating the timing and upcoming products and services the project will deliver. Which software should she use to manage the website?

A. Electronic document and records management system (EDRMS)

B. Content management system (CMS)

C. Customer relationship management (CRM)

D. Enterprise resource planning (ERP)

182. Software development kits (SDKs), user guides, run books, and product manuals are all examples of which of the following?

A. Data warehouse

B. Customer relationship management

C. Documentation

D. Content management systems

183. The project team has been asked to turn over all relevant records regarding the project and to assign the documents a category determining how long to keep the documentations after the project is completed. Where should the project manager look to evaluate which categories should be used?

A. Organizational retention schedule

B. Project charter

C. Project management information systems

D. Data warehouse

184. A local government organization has determined the cost of hosting a disaster recovery site is prohibitive and so they are seeking a service provider capable of meeting their recovery time and recovery point objectives. What type of cloud model is this agency looking to use?

A. Software as a service (SaaS)

B. Platform as a service (PaaS)

C. Anything as a service (XaaS)

D. Infrastructure as a service (IaaS)

185. With the adoption of the European Union's General Data Protection Regulation (GDPR) and the California Consumer Privacy Act (CCPA), which organizational business unit becomes a more important stakeholder for projects?

   A. Legal

   B. Public relations

   C. Finance

   D. Information technology

186. Which privacy consideration law requires integrating specific data protection practices into business offerings as well as adding processes to address data subject requests?

   A. GDPR

   B. CCPS

   C. HIPPA

   D. All of the above

   E. None of the above

187. Which discipline helps legal departments stay in compliance with emerging laws on an ongoing basis?

   A. Project management

   B. Asset management

   C. Change management

   D. Internal and external audit

188. All of the following are reasons project management needs to be involved to achieve regulatory compliance EXCEPT:

   A. Regulations change over time.

   B. Data is being collected on an exponential pace.

   C. Laws of other countries do not carry regulatory impact.

   D. New regulations are constantly being enacted.

189. All of the following are risks of noncompliance of country, state, or province-specific privacy regulations in project management EXCEPT:

   A. Increase in security breaches

   B. Increase in consumer privacy considerations

   C. Loss of productivity

   D. Reputational damage

190. A steering committee is questioning a proposed budget item in a project budget to ensure compliance with privacy laws. The committee debates the need for this activity to be completed as it carries a high cost. What is the best argument the project manager can make for the investment in compliance?

A. Compliance is the cost of doing business.

B. Noncompliance saves the company money and can be struck.

C. No one will find out if the project does not perform that activity.

D. Noncompliance costs the company more money.

191. Which of the following are not risks of failing to comply to regulatory requirements? (Choose two.)

A. Penalties and fines

B. Gain a competitive advantage

C. Compliance litigation

D. Reputational damage

E. Restricted access to markets and product delays

F. Regulated out of business

192. What is a project manager's responsibility to behave ethically?

A. Project managers abide by a code of ethics.

B. It depends on the organization.

C. Project managers must focus on schedule, budget, and scope above all things.

D. Ethical behavior does not apply to project management.

193. Professionals who act in an honest, responsible, and respectful way are exhibiting which type of behavior?

A. Unethical

B. Criminal

C. Ethical

D. Regulatory

194. In which project life cycle phase should the regulatory impact assessment process begin?

A. Discovery/concept preparation

B. Initiation

C. Planning

D. Execution

E. Closing

195. All of the following would be considered basics for compliance in project management EXCEPT:

    A. Add compliance activities to the project schedule.

    B. Ensure regulatory management responsibility is assigned.

    C. Monitor changes in regulation.

    D. Manage compliance risk and activities as its own project.

196. Wigit Construction has been hired to create a new highway road signage system for a large state. They have undertaken this project in other states as well, so they have a solid boiler-plate for their project activities. Does a regulatory scan still make sense for this project due to repeatability of successful projects delivered in other states?

    A. No, highway and constructions projects are beholden to national laws and regulations, so compliance is assured.

    B. Yes, state-level regulations may differ, and a regulatory scan will help to illuminate the requirements.

    C. No, highway projects are not concerned with privacy implications as it serves a public good.

    D. Yes, even as a non-value-added activity, the documentation of performing the step will reduce future liability.

197. In 2015, the U.S. federal government enacted the Program Management Improvement and Accountability Act. What area of compliance is this area focused on?

    A. Privacy

    B. Project management

    C. Financial management

    D. Healthcare

198. A company utilizing its offshore team has just won a contract to work on replacing a government agency's financial system. The contract stipulates there is personally identifiable information stored in the system and has a service-level agreement enforcing this clause. What steps if any should the project manager take as the project team is built?

    A. Educate the team on the importance of cybersecurity.

    B. Educate the team on the contract clause and their responsibilities.

    C. Educate the team on personally identifiable data and how to handle it.

    D. All of the above.

    E. Only A and C.

199. Benched resources are when the project has which of the following issues?

    A. Individuals who are finished with the project but have not yet started a new assignment

    B. Individuals who have too much work for them to be able to complete the project

    C. A lack of talent in the industry, which leads to a shortage of qualified personnel on the project

    D. Individuals ordered to the sidelines because of their performance

200. In which of the following situations would team-building efforts provide the most impact on a project? (Choose three.)

   A. Team discord

   B. Schedule changes

   C. Missed deliverables

   D. Project phase completion

   E. Lessons learned meeting

   F. Change in project manager

201. Which of the following statements is true regarding a SWOT analysis?

   A. Strengths/weaknesses are external to the organization; opportunities/threats are internal to the organization.

   B. Strengths/weaknesses/opportunities/threats are all external to the organization.

   C. Strengths/weaknesses are internal to the organization; opportunities/threats are external to the organization.

   D. Strengths/weaknesses/opportunities/threats are all internal to the organization.

202. In the project management context, what does COQ stand for?

   A. Cost of quality

   B. Critical to quality

   C. Cost of quantities

   D. Critical of quantities

203. Which of the following would you need in the calculation of the risk score? (Choose two.)

   A. SPI

   B. Risk impact

   C. CPI

   D. Risk probability

   E. Risk trigger scores

   F. EAC

204. What is the critical chain method?

   A. Schedule network analysis technique

   B. Dependency model

   C. Signature path for project charter approval

   D. Earned value method

205. Harry works for a research and development firm trying to create a revolutionary new product. The likely risk tolerance for the company is which one of the following?

   A. Very high

   B. High

   C. Low

   D. Very low

Chapter

5

Practice Test 1

1. What are two defining characteristics of a project? (Choose two.)
   A. Reworking an existing project
   B. An organized effort to fulfill a purpose
   C. Has a specific end date
   D. Routine activities to an organization
   E. Blueprints needed to construct a building

2. Why would a buyer issue a purchase order?
   A. To describe the quantity of goods and services needed and what price will be paid.
   B. To outline the intent or actions of both parties before entering a contract.
   C. To stop another company from using their intellectual property and to inform them not to do it again.
   D. The situation requires more flexibility and simplicity than a contract.

3. To which project-centric document would you refer to get a description of the project, find out the key deliverables, and gain an understanding of the success and acceptance criteria for the project?
   A. Project charter
   B. Project schedule
   C. Scope statement
   D. Meeting minutes

4. At the completion of a government project requiring security clearance, what document would a vendor likely be asked to sign?
   A. Nondisclosure agreement
   B. Service-level agreement
   C. Warranty
   D. Request for proposal

5. Which of the following data sources are considered SPII? (Choose two.)
   A. Social Security number (SSN)
   B. Political party
   C. First name
   D. City
   E. Driver's license number

6. When the schedule slips on a project because the work is taking longer than planned, what type of common project change does this represent?
   A. Funding change
   B. Timeline change
   C. Risk event
   D. Requirements change

7. The process of assigning numerical probabilities to each risk and the impacts on project objectives is known as which one of the following?

   A. Identify risk

   B. Quantitative risk analysis

   C. Qualitative risk analysis

   D. Risk categorization

8. What type of project management tool is depicted here?

Task Name	Duration	Start	Finish	4/17	4/24	May 5/1
⊿ **Planning**	**11 days?**	**Fri 4/22/22**	**Fri 5/6/22**			
Assess resource pool	3 days	Fri 4/22/22	Tue 4/26/22			
Train project team members	1 day	Wed 4/27/22	Wed 4/27/22			
Develop communication plan	1 day	Wed 4/27/22	Wed 4/27/22			
Develop a detailed scope statement	1 day?	Thu 4/28/22	Thu 4/28/22			
Define units of work	1 day?	Fri 4/29/22	Fri 4/29/22			
Develop project schedule	5 days	Mon 5/2/22	Fri 5/6/22			
Determine budget considerations	1 day?	Mon 5/2/22	Mon 5/2/22			
Develop QA plan	1 day?	Mon 5/2/22	Mon 5/2/22			
Perform initial risk assessment	1 day?	Mon 5/2/22	Mon 5/2/22			
Develop tranistion plan	1 day?	Tue 5/3/22	Tue 5/3/22			
Develop project plan	1 day?	Wed 5/4/22	Wed 5/4/22			5/4
▹ **Execution**	**15.13 days**	**Thu 5/5/22**	**Thu 5/26/22**			

   A. Fishbone diagram

   B. Gantt chart

   C. Collaboration tools

   D. Process diagram

9. Which project role promotes the need and urgency of a project, as well as advertising the project's success?

   A. Project scheduler

   B. Project manager

   C. Project sponsor or champion

   D. Project coordinator

10. The EV for a project is 800, and the AC is 1200. The CPI for the project would be which one of the following?

    A. 1.5

    B. 400

    C. 2000

    D. .667

11. As a part of a daily Scrum meeting, the team seeks to share information in three areas. Which set of questions represents the topics of this knowledge sharing?

    A. What did I accomplish today? Who will you be working with today? What obstacles are preventing progress?

    B. What did I accomplish yesterday? What will I do today? What are the necessary next steps?

    C. What did I accomplish yesterday? What will I do today? What obstacles are preventing progress?

    D. What did I accomplish yesterday? Who will you be working with today? What obstacles are preventing progress?

12. Which document describes the goods or services an organization is interested in procuring from outside the organization?

    A. SOW

    B. RFI

    C. RFP

    D. RFQ

13. What is a software application that manages back-office activities such as accounting, human resources, supply chain, operations, and procurement called?

    A. Database management system (DBMS)

    B. Enterprise resource planning (ERP)

    C. Content management system (CMS)

    D. Customer relationship management (CRM)

14. What does SWOT stand for?

    A. Situation, work, open source, traceability

    B. Strengths, work, opportunities, traceability

    C. Situation, weaknesses, open source, threats

    D. Strengths, weaknesses, opportunities, threats

15. Cheryl is a technical lead on a project that is wrapping up remote work at a customer site. What is the best method to communicate the work efforts and next steps with the customer?

    A. Closure meeting

    B. Virtual meeting

    C. Kickoff meeting

    D. In-person meeting

16. Which cloud model is a software licensing and delivery model where software is licensed on a subscription basis and is centrally hosted?

A. XaaS

B. PaaS

C. SaaS

D. IaaS

17. What does MFA stand for?

A. Multifactor architecture

B. Multifactor authentication

C. Mid-tier form architecture

D. Middle factor authentication

18. The approved budget for given work to be completed within a specific timeframe is called which one of the following?

A. Cost variance

B. Planned value

C. Earned value

D. Actual costs

19. A change manager supports a large transportation agency. In support of an upcoming project change, she has established emails to be sent two weeks in advance, a week in advance, and the day of the changes so that users can expect an outage and a new layout when the service comes back online. What is she working on?

A. Maintenance windows

B. Rollback plan

C. Validation checks

D. Notifications

20. Which law established protected health information and the rules governing its release?

A. HIPPA

B. OSHA

C. DoCRA

D. PgMP

21. A company is considering opening a store in a new country in which it has never done business before. This new country has laws called the consumer bill of rights that outline a company's responsibility to collect private data as well as a consumer's ability to see and/or restrict the collection of this data. Which ESG factor would influence this project's activities?

A. Project impact to the local and global environment

B. Awareness of company vision, mission statements, and values

C. Awareness of appliable regulations and standards

D. Project impact to company brand value

22. Developing contingency reserves to deal with risks, should they occur, is known as which one of the following?

    A. Active acceptance

    B. Passive acceptance

    C. Exploit

    D. Transfer

23. The project team has been asked to include a data validation module with the software they are developing, and this will add an additional three weeks to the project. This is an example of which one of the following?

    A. Issue

    B. Risk

    C. Change

    D. Milestone

24. What are milestones?

    A. Checkpoints on a project to determine Go/No-Go decisions

    B. A measure of the distance traveled on a project

    C. Characteristics of deliverables that must be met

    D. Major events in a project used to measure progress

25. What type of project management tool is depicted here?

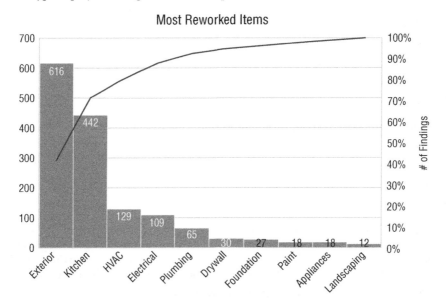

A. Pareto chart

B. Project schedule

C. Histogram

D. Gantt chart

26. At the beginning of a work period, or sprint, there is a sprint planning meeting. What does this meeting accomplish?

   A. Sets a realistic backlog of items to be completed during this iteration

   B. Sets the communication and quality plans for the project

   C. Prepares the project charter and kickoff meeting

   D. Gets a head start on the work needed on the project

27. Which vendor solicitation method is best to use when more data is needed about the goods and services that need to be procured?

   A. Request for proposal (RFP)

   B. Request for quotation (RFQ)

   C. Request for information (RFI)

   D. Statement of work (SOW)

28. The Payment Card Industry Data Security Standards (PCI DSS) address the handling and management of payment card data. It covers all aspects of payment card data handling, including acquiring, transmitting, storing, and processing the data. What is the PCI DSS an example of?

   A. Personally identifiable information

   B. Industry-specific compliance

   C. Regulatory impacts

   D. Country-specific privacy stipulation

29. A systematic and independent examination of project procedures, documentation, spending, statutory compliance, and reporting is known as which one of the following?

   A. Audit

   B. Gate review

   C. Scrum retrospective

   D. Project change

30. A team is being assembled for a financial company that looks for the highest standards in ethical behavior. The project manager is excited by a candidate's skills, attitude, and experience, but a background check reveals the candidate misrepresented their education and work experience. Which ESG factor likely influenced the decision to move away from this candidate?

   A. Project impact to the local and global environment

   B. Awareness of applicable regulations and standards

   C. Awareness of company vision, mission statements, and values

   D. Project impact to company brand value

31. All of the following are examples of deliverables EXCEPT:

    A. Blueprints

    B. User documentation

    C. The finished product

    D. Sign-off on the project charter

32. Which of the following tools is used for entering data to generate a Gantt chart, WBS, or activity sequence automatically?

    A. Process diagram

    B. Scrum retrospective

    C. Collaboration tools

    D. Project scheduling software

33. What is a special form of a bar chart that visually displays the central tendency, dispersion, and distribution for statistical data?

    A. Kanban board

    B. Histogram

    C. Pareto chart

    D. SIPOC-R

34. Once a change request is submitted, where should it be recorded and assigned an identification number for tracking purposes?

    A. Change request log

    B. Risk register

    C. Business process repository

    D. Issue log

35. In which project life cycle phase would the issues log generally be developed?

    A. Discovery/concept preparation

    B. Initiation

    C. Planning

    D. Execution

    E. Closing

36. A meeting has just been conducted on a project where the expectations, goals, and objectives of the project have been explained, including the milestones and timelines of the project. Project sign-off is also likely to occur during this meeting. What type of meeting was this?

    A. Lessons learned

    B. Kickoff meeting

    C. Status meeting

    D. Gate check

37. What is the project artifact that breaks down resources by category and type?

    A. Resource breakdown structure

    B. Organizational breakdown structure

    C. Equipment breakdown structure

    D. Work breakdown structure

38. Liberty in purchasing needs you to create an SOW prior to the releasing an RFP. What does SOW stand for?

    A. Service of workforce

    B. Statement of work

    C. Statement of workforce

    D. Support of work

39. Your friend has hinted that she would like a special gift for her birthday. The following steps are needed for the gift exchange:

Enjoyment of her gift.

Put card on the gift.

Wrap present.

Choose gift.

Give her the present.

Make purchase.

Fill out the card.

What is the correct sequence for project Happy Birthday?

    A. Choose gift; make purchase; put card on gift; wrap present; fill out card; enjoyment of her gift; give her the present.

    B. Give her the present; put card on gift; wrap present; make purchase; fill out card; enjoyment of her gift; choose gift.

    C. Put card on gift; give her the present; fill out card; make purchase; enjoyment of her gift; choose gift; wrap present.

    D. Choose gift; make purchase; wrap present; fill out card; put card on gift; give her the present; enjoyment of her gift.

40. A vendor is very strict on insisting on formal change control on a project for a customer and few changes are actually getting approved. Which contract vehicle would cause a vendor to behave in this way?

    A. Fixed-price

    B. Request for proposal

    C. Time and materials

    D. Cost-reimbursable

41. The project team is required to attend a daily stand-up meeting to discuss the project activities and roadblocks. Which document lets the project team know that they need to attend this meeting?

A. Project schedule

B. Scope statement

C. Communication plan

D. Project charter

42. The joining of two businesses to come together and operate as one, single entity is known as which one of the following?

A. Business acquisition

B. Business merger

C. Business process change

D. Business split

43. What does the acronym RACI stand for?

A. Responsible, accountable, consulted, and informed

B. Responsibility, authority, consult, and inform

C. Responsible, authority, consulted, and inform

D. Responsibility, accountable, consult, and informed

44. What are the three common constraints found in projects? (Choose three.)

A. Inventory

B. Personnel

C. Environment

D. Time

E. Budget

F. Scope

45. A government agency has hired a firm to perform work on its sewer system. As a part of the RFP, the agency requires the successful vendor to carry insurance for errors and omissions. What type of risk strategy is this?

A. Transfer, negative risk strategy

B. Exploit, positive risk strategy

C. Accept, negative risk strategy

D. Share, positive risk strategy

46. In what stage of team development do team members begin to confront each other and vie for position and control?

A. Adjourning

B. Norming

C. Forming

D. Storming

E. Performing

47. In which project life cycle phase would the brainstorming, evaluation, and impact of risk be assessed?

A. Discovery/concept preparation

B. Initiation

C. Planning

D. Execution

E. Closing

48. What elements are explained in a business case?

A. Justification by identifying the organizational benefits

B. Alternative solutions

C. Alignment to the strategic plan

D. All of the above

E. A and C

49. In what project life cycle phase is the influence of stakeholders the least effective?

A. Discovery/concept preparation

B. Initiation

C. Planning

D. Execution

E. Closing

50. When is a project considered to be a success?

A. Stakeholder expectations have been met.

B. The phase completion has been approved.

C. All project phases have been completed.

D. The vendor has been released from the project.

51. Which tool would you use to create a visual representation of timelines, start dates, durations, and activity sequence?

A. Histogram

B. Gantt chart

C. Process diagram

D. Pareto chart

52. What is the comprehensive collection of documents that spells out communication, risk management, project schedule, and scope management?

A. Request for proposal

B. Project management plan

C. Dashboard information

D. Scope statement

53. What type of agreement should organizations use to ensure their trade secrets are not revealed or used by other companies?

A. NDA

B. MOU

C. RFP

D. SLA

54. What is the purpose of a data warehouse?

A. To handle extremely large data environments

B. To separate the data's physical location from other applications

C. To bring various data sources together for the purpose of analysis and decision-making

D. To bring all the company's unstructured data together in a central location

55. A project is seeking a third party to handle hardware and software tools for the application development team to complete a project. Which cloud model should the project team look to use?

A. SaaS

B. IaaS

C. XaaS

D. PaaS

56. In a multitiered architecture, which tier is where information is stored, sorted, and indexed?

A. Data

B. Processing

C. Application

D. Presentation

57. A new U.S. state law gives customers the ability to request what data a company is storing about their users and how the company is using this information. A global company is having to adjust their product design to conform with this requirement. Which ESG factor can cause the project to change in this manner?

A. Awareness of applicable regulations and standards

B. Awareness of company vision, mission statements, and values

C. Project impact to company brand value

D. Project impact to the local and global environment

58. A list of steps undertaken to undo a release and restore a system to its original state is known as what?

A. Risk assessment

B. Anything as a service

C. Continuous integration process

D. Rollback plan

59. What is a licensed professional who oversees the design of a project known as?

A. Subject matter expert

B. Scrum master

C. Architect

D. Project manager

60. During a regular check-in meeting with stakeholders, the project manager presents information on all task areas. Each task area has a column for time, cost, resources, risks, issues, and changes, and each item shows a red, yellow, or green indication representing the condition of each task in each area. What project-centric document is being presented?

A. Project dashboard

B. Gantt chart

C. Process diagram

D. Run chart

61. What kind of tool would automate the creation of critical path, float, WBS, and activity sequence?

A. Project management plan

B. Project scheduling software

C. PERT

D. Run chart

62. Which tool would be used to display observed data in a time sequence?

A. Pareto chart

B. Histogram

C. Run chart

D. Scatter diagram

63. Which knowledge management tool would be used for a team to communicate instantly, share information on task ownership and status, and see events for an entire work group?

A. Wiki pages

B. Intranet sites

C. Vendor knowledge bases

D. Collaboration tools

64. A subject matter expert on a project has been monitoring the distribution of support calls received each hour during the day to determine staffing needs for a project. Which tool is this person most likely using?

A. Gantt chart

B. Histogram

C. Run chart

D. Process diagram

65. After establishing the product backlog, what tool would you use to determine the project's velocity?

    A. Burndown chart

    B. Kanban board

    C. Fishbone diagram

    D. Sprint speed

66. What type of project management tool is depicted here?

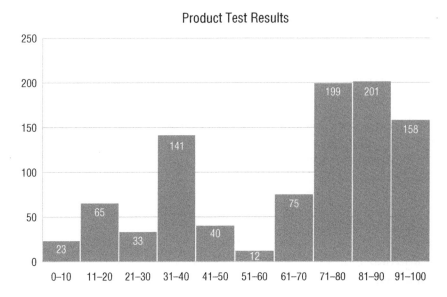

A. Process diagram

B. Histogram

C. Pareto chart

D. Gantt chart

67. Which office productivity tool would be best to record spending so far on a project so that the project team can keep track of where money has been spent and how much budget is remaining?

    A. Word processing

    B. Spreadsheets

    C. Presentation

    D. Charting/diagramming

68. What are the two forms of acceptance when considering risk response strategies?

  A. Unintentional

  B. Active

  C. Deliberate

  D. Passive

69. The Closing phase includes all of the following EXCEPT:

  A. Review of lessons learned

  B. Release of project members

  C. Archiving of project documents

  D. Monitoring the risks and issues log

70. What is the process used to document, identify, and authorize changes in an IT environment?

  A. Change control

  B. Executive oversight

  C. Organizational change management

  D. Promote to production

71. The project team has been asked to add some additional functionality to an existing application. A business analyst sits down with the stakeholders to be walked through what the application needs to do, how it needs to look, and what other applications or reports the data needs to be shared with. What part of the change control process is the business analyst performing?

  A. Customer notifications

  B. Requirements definition

  C. Release management

  D. Rollback planning

72. At a gate check meeting, the committee is presented with a report that shows health in different areas of a project: Schedule = Green, Budget = Yellow, Scope = Green, Risk = Yellow, and Publicity = Red. What type of tool is the committee looking at?

  A. Pareto chart

  B. Scatter chart

  C. Collaboration tool

  D. Balanced scorecard

73. An animated movie project requires the following ordered steps: story idea, script, storyboard, voice talent, models, and sets construction, to name just a few. Storyboard has what relationship to script?

  A. It is a predecessor task.

  B. It is a discretionary task.

  C. It is a successor task.

  D. It is a mandatory task.

74. All of the following are cost-estimating techniques EXCEPT:

    A. SIPOC-R

    B. Bottom-up estimating

    C. Parametric estimating

    D. Analogous estimating

75. The work breakdown structure is created during which project life cycle phase?

    A. Discovery/concept preparation

    B. Initiation

    C. Planning

    D. Execution

    E. Closing

76. What is the indication of how fast a project is spending through its budgeted money?

    A. Planned value

    B. Burn rate

    C. Cost variance

    D. Expenditure tracking

77. Which of the following is the best definition of a capital expense, or CapEx?

    A. Money spent on running the business operations of a company

    B. Money spent on long-term physical or fixed assess used in a business's operations

    C. Debt accrued on travel expenses needed to run a business

    D. Debt accrued through the issuing of long-term bonds

78. Which of the following is the best definition of an operational expense, or OpEx?

    A. Money spent on long-term physical or fixed assess used in a business's operations

    B. Debt accrued through the issuing of long-term bonds

    C. Money spent on running the business operations of a company

    D. Debt accrued on travel expenses needed to run a business

79. A seller that has passed through a first phase of onboarding through evaluation using a standard criterion to determine a seller's ability to supply a good or service is known as what?

    A. Bidder's conference

    B. Prequalified vendor

    C. Terms of reference

    D. Statement of work

80. The characteristics of the lower-level WBS include all of the following EXCEPT:

   A. WBS components are a further decomposition of project deliverables.

   B. WBS components should always happen concurrently with determining major deliverables.

   C. WBS components should be tangible and verifiable.

   D. WBS components should be organized in terms of project organization.

81. When breaking down project deliverables, what is the lowest level that is recorded in a WBS?

   A. Daily work schedules

   B. High-level requirements

   C. Work package

   D. Major milestones

82. What type of project management tool is depicted here?

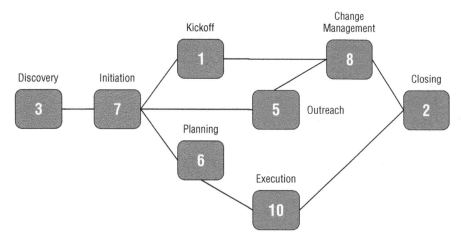

   A. Program Evaluation Review Technique (PERT) chart

   B. Gantt chart

   C. Project dashboard

   D. Budget burndown chart

83. In which project life cycle phase would you develop a transition or release plan?

   A. Discovery/concept preparation

   B. Initiation

   C. Planning

   D. Execution

   E. Closing

84. A Sprint backlog is composed of which of the following? (Choose three.)

    A. Action plan

    B. Gate review

    C. Run charts

    D. Product backlog items

    E. Milestone chart

    F. Sprint goal

85. Why is it important to validate deliverables when a project starts to conclude?

    A. To confirm the project has satisfied the objectives

    B. To accept deliverables to the scope

    C. To verify timeliness and completeness to pay a vendor

    D. All of the above

    E. A and C

86. As a project concludes, the project manager meets with the project team to review expenditures for the project and signs off on the project budget. This is known as which of the following?

    A. Return on investment analysis

    B. Budget reconciliation

    C. Current state analysis

    D. Updating the project budget

87. The project charter is prepared and agreed to in which project life cycle phase?

    A. Discovery/concept preparation

    B. Initiation

    C. Planning

    D. Execution

    E. Closing

88. A team has set guidance that project communication can only take place through the company-issued collaboration platform or through company-issued devices. This is an example of which of the following?

    A. Identify and assess stakeholders.

    B. A responsibility assignment matrix.

    C. Establish accepted communication channels.

    D. Establish communication cadence.

89. What type of communication method would make sense for routine status meetings on a project where the team is spread out to different cities on the same continent?

   A. In-person meetings

   B. Virtual meetings

   C. Closure meetings

   D. Kickoff meetings

90. A construction company is in the middle of a project to build a bridge. The earned value for the project is $9,500, and the actual cost for the project is $7,000. Select the correct cost variance for the project and its meaning:

   A. $2,500 and the project is under budget

   B. $–2,500 and the project is under budget

   C. $2,500 and the project is over budget

   D. $–2,500 and the project is over budget

91. All electronic communications and meetings notes are stored with the project file at the conclusion of the project. This is an example of:

   A. Communication security

   B. Communication integrity

   C. Communication archiving

   D. Communication planning

92. What type of project management tool is depicted here?

Req ID	Requirement description	Business justification	Status	Comments
1	Add business intelligence to dashboard	Will enable real time visibility to accident data	Done	Dec 28: started Jan 8: Defect reported Jan 30: Defect fixed
2	Create interface to finance data	Will enable managers to see budget and expenditures	In progress	Make sure to adjust for local currency
3	Add mobile view to web page	Will allow team members to view data in the field	Hold	May not be possible with current project budget
4	—	—	—	—

   A. Fishbone/Ishikawa diagram

   B. Burndown chart

   C. Requirements traceability matrix

   D. Run chart

93. In the development of a project schedule, setting governance gates are important. All of the following are examples of governance gates EXCEPT:

   A. Daily stand-up meetings

   B. Client sign-off

   C. Management approval

   D. Legislative approval

94. When a meeting agenda sets a fixed time on a schedule where a task must be completed within that time, what technique is being used?

   A. Soft timeboxing

   B. Hard timeboxing

   C. Brainstorming

   D. Action items

95. Total expenditures for completed project work within a specific timeframe is known as which one of the following?

   A. Earned Value

   B. Planned Value

   C. Cost Variance

   D. Actual Cost

Chapter

6

Practice Test 2

1. Which project management role is responsible for product integrity and ensures conformance to standards and specifications?

    A. Developer/engineers

    B. Business analyst

    C. Testers/quality assurance specialists

    D. Architect

2. Which of the following is *not* a communication trigger?

    A. Project planning

    B. Kickoff meeting

    C. Milestones

    D. Schedule changes

3. A potential future event that can have either a negative or positive impact on a project is known as which one of the following?

    A. A risk

    B. A deliverable

    C. A requirement

    D. An issue

4. What risk tolerance would a start-up in the technology industry likely have?

    A. Risk taker

    B. Risk avoider

    C. Gambler

    D. Moderate risk tolerance

5. As part of the preparation to begin work on a project, the project manager makes a list: (1) There can be no work on weekends and holidays, (2) overtime is not approved without steering committee consent, and (3) seventy-five percent of the project team must be union staff. This list represents which of the following?

    A. Objectives

    B. Requirements

    C. Constraints

    D. Assumptions

6. What does it mean to exploit a positive risk?

    A. Assign the risk to a third party who is best able to bring about opportunity.

    B. Monitor the probability or impact of the risk event to ensure benefits are realized.

    C. Choose to accept the consequences of the risk.

    D. Look for opportunities to take advantage of positive impacts.

7. A start-up company is constructing a new mobile home sharing app. Users of the service will be required to share their location data to be able to sign up for a trip using someone else's mobile home. Which of the following ESG factors would the company need to consider for this project?

   A. Mitigating risk factors

   B. National and local privacy laws and regulations

   C. Critical path activities impacting project planning

   D. Choosing waterfall versus an Agile approach

8. Which authentication method requires the user to provide two or more pieces of verification evidence to log into their account?

   A. Certificate-based authentication

   B. Biometric authentication

   C. Multifactor authentication

   D. Token-based authentication

9. DewDrops has gone through a rapid growth cycle and now has several one-off systems for activities like the company's finances, supply, and human resources. The leaders of the company want to have a system that brings these activities together. What kind of system would meet these needs?

   A. ERP

   B. XaaS

   C. LAN

   D. CRM

10. What is a user story in the Agile methodology?

   A. Customer survey results after product release

   B. Key information about stakeholders and their jobs

   C. A high-level definition of someone using the product or service

   D. A visual representation of product burndown

11. What is the single role in a RACI matrix that can only be assigned once per task?

   A. Consulted

   B. Responsible

   C. Informed

   D. Accountable

12. Which of the following would not receive communication if there was an update to the risk register?

   A. Project sponsor

   B. Project team

   C. PMO

   D. Project manager

13. Dyan is applying for a job working on a defense project for a government agency. To help confirm her employment, what steps might be required for Dyan to be hired as the project manager? (Choose two.)

A. Service-level agreement

B. Background screening

C. Clearance requirements

D. Branding restrictions

14. A movie production facility has a secure door that can only be accessed through a retinal scan and an authorized badge being scanned. What security concept is being enacted in this scenario?

A. Branding restrictions

B. Scrum retrospective

C. Legal and regulatory impacts

D. Facility access

15. Once a change request is submitted, where should it be recorded and assigned an identification number for tracking purposes?

A. Risk register

B. Business process repository

C. Issue log

D. Change request log

16. The project charter has a filename of `Projectcharter1`. The project manager makes an update to the file, and it now is called `Projectcharter2`. This is an example of which one of the following?

A. Communicating changes

B. Version control

C. Change request logs

D. Risk register

17. What type of project management tool is depicted here?

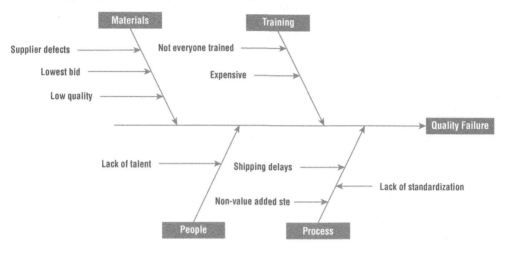

A. Process diagram

B. Histogram

C. Pareto chart

D. Fishbone diagram

18. An airline engineering company is forced to cut the project budget after poor financial results in the previous quarter. How would this most likely impact the project?

A. The project will take longer because the number of resources is cut.

B. The project team goes to the steering committee for more funds.

C. The project scope is cut back to operate within the new budget.

D. The project is postponed due to lack of financial resources.

19. A resource shortage means which one of the following?

A. There are too many resources, leading to underallocation.

B. There is a shortage of things for team members to work on.

C. There are not enough resources for the task, leading to overallocation.

D. There is an abundance of things for team members to work on.

20. Which of the following are ways to organize the WBS? (Choose three.)

A. Backlog and sprints

B. Project phases

C. Subprojects

D. Major deliverables

E. Prioritized by risk

21. A project team tasked with creating a new software application has discovered that the programming needed is twice as complex as they thought it would be, and they have informed the project manager that they will need more resources or will need to push the due date out by two months. Where would this topic get recorded in project documentation?

    A. Project charter

    B. Issue log

    C. Action items

    D. Risk register

22. A project sponsor or champion serves what role on a project?

    A. Develops and maintains the schedule

    B. Is the approval authority and removes roadblocks

    C. Sets the standards and practices for projects in the organization

    D. Performs cross-functional coordination

23. Which type of contract allows the seller to recover all allowable expenses associated with providing the goods or services?

    A. Time and materials

    B. Fixed-price contract

    C. Statements of work

    D. Cost-reimbursable contract

24. The comprehensive collection of documents that spells out communication, risk management, project schedule, and scope management is known as which one of the following?

    A. Request for proposal

    B. Dashboard information

    C. Project management plan

    D. Scope statement

25. Project team members are required to report to both the project manager and their functional manager, who share authority for the resources. What type of organization is this?

    A. Weak matrix

    B. Balanced matrix

    C. Projectized

    D. Strong matrix

26. What type of project management tool is depicted here?

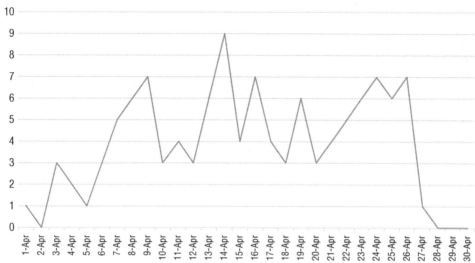

Dropped Calls on New Phone Switch

A. Run chart

B. Histogram

C. Pareto chart

D. Fishbone diagram

27. Which tool would be used to help find the distribution of issues (or other variable) from highest to lowest as bars on a chart?

A. Scatter chart

B. Histogram

C. Pareto chart

D. Run chart

28. What are major accomplishments of the project or key events known as?

A. Requirements

B. Deliverables

C. Milestones

D. Risks

29. The situation where the project team is stuck on the last piece of work, which prevents the project from completing, is known as which one of the following?

    A. Pareto diagram

    B. The 95 percent phenomenon

    C. IRR

    D. The 80/20 rule

30. Assessing the status of the budget, reviewing risks/issues logs, and measuring performance occurs in what project life cycle phase?

    A. Discovery/concept preparation

    B. Initiation

    C. Planning

    D. Execution

    E. Closing

31. All of the following are examples of categories that might be included in a resource breakdown structure EXCEPT:

    A. External

    B. Project management

    C. Organizational

    D. Probability and impact

32. Mark submits a purchase requisition to his company's procurement section. He must wait for a PO to be cut in order for the vendor to begin work. What is a PO?

    A. Planned objective

    B. Purchase office

    C. Purchase order

    D. Planned order

33. A project manager is having problems with one team member who is being insubordinate. The project manager approaches the team member and communicates how the behavior is inappropriate and how the team member will behave from now on. This is an example of which one of the following?

    A. Forcing

    B. Avoiding

    C. Confronting

    D. Smoothing

34. What are governance gates?

A. A checkpoint between project phases where approval is obtained to move forward

B. A checkpoint where quality is checked against a previously established criterion

C. Checkpoints at the beginning and end of the project only

D. After a project governor is appointed, unplanned interruptions from this project sponsor

35. Even though the project scope statement has been approved, the customer has routinely asked for more features to be added to the product, causing the due date and resources to be adjusted consistently. This is an example of which type of influence?

A. Change request

B. Constraint reprioritization

C. Schedule constraint

D. Scope creep

36. After establishing the product backlog, what tool would you use to determine the project's velocity?

A. Fishbone diagram

B. Burndown chart

C. Kanban board

D. Sprint speed

37. How should a project team record what was said and agreed to during project gatherings?

A. Create a scope statement.

B. Create a business case.

C. Create meeting minutes.

D. Create a project management plan.

38. Which risk response strategy attempts to minimize the impact or the probability of a negative risk?

A. Mitigate

B. Accept

C. Transfer

D. Avoid

39. At the conclusion of a defined work period, Belle got the project team together to figure out what went well, what did not go well, and what improvements could be made.

What type of meeting was this?

A. Scrum retrospective

B. Product backlog

C. Daily Scrum

D. Kickoff meeting

40. In what project life cycle phase are the majority of the processes and project documents created?

   A. Discovery/concept preparation

   B. Initiation

   C. Planning

   D. Execution

   E. Closing

41. Chase is a project manager, and they have released all of the team members from the project, closed vendor contracts, and archived project documents. In what life cycle phase is the project?

   A. Discovery/concept preparation

   B. Initiation

   C. Planning

   D. Execution

   E. Closing

42. Resource allocation, including assigned equipment, team members, and money to support a project, occurs in which project phase?

   A. Discovery/concept preparation

   B. Initiation

   C. Planning

   D. Execution

   E. Closing

43. Teams normally go through a similar development cycle. Which is the correct order of those stages?

   A. Forming, norming, performing, storming, and adjourning

   B. Norming, storming, forming, adjourning, and performing

   C. Norming, forming, storming, adjourning, and performing

   D. Forming, storming, norming, performing, and adjourning

44. The project team is ready to make a change to a web application already in use by an organization. Which artifact should the project manager refer to in rolling out the changes?

   A. Tiered architecture

   B. Downtime/maintenance window schedules

   C. Data warehouse

   D. Customer notifications

45. What are the three common constraints found in projects? (Choose three.)

    A. Scope

    B. Personnel

    C. Working space

    D. Quality

    E. Budget

    F. Time

46. Which of the following is both a serious risk management issue and a compliance/legal matter?

    A. Branding restrictions

    B. Anything as a service

    C. Data security classification

    D. Data privacy

47. Which information security concept addresses where and how your data is stored?

    A. Application security

    B. Digital security

    C. Network and system security

    D. Physical security

48. The commitment to keep a person's or organization's information private and protect unauthorized disclosure of information from being disclosed is known as which of the following?

    A. Confidentiality

    B. Data sensitivity

    C. Privacy

    D. Security clearance

49. Which project management tool helps determine whether a system, process, or program involving personal information raises privacy risks?

    A. Multitiered architecture

    B. Privacy impact assessment

    C. Service-level agreement

    D. Backlog prioritization

50. What is the definition of a work breakdown structure?

    A. A deliverable-oriented decomposition of a project

    B. A graphic representation of tasks and their sequence

    C. A high-level outline of milestones on a project

    D. A task-oriented decomposition of a project

51. What are project requirements?

    A. A measure of the distance traveled on a project

    B. Major events in a project used to measure progress

    C. Checkpoints on a project to determine Go/No-Go

    D. Characteristics of deliverables that must be met

52. Expenditures not expensed directly on a company's income statement and that are considered an investment are known as which of the following?

    A. SOW

    B. OpEx

    C. TOR

    D. CapEx

53. What does TOR stand for?

    A. Time of reflection

    B. Terms of reconciliation

    C. Terms of reference

    D. Time of reference

54. The calculation of the rate of return for a given investment for a given period of time is known as which of the following?

    A. Payback period

    B. Return on investment

    C. Time value of money

    D. Earned value

55. During a project meeting, a key stakeholder looks at a report and sees that the schedule now has a yellow setting when last week it was set at green. What kind of data is the stakeholder viewing?

    A. Dashboard information

    B. Project management plan

    C. Ishikawa diagram

    D. Risk register

56. A team that is geographically dispersed has the need to work on documents together and to have more tools than phone calls by which they can communicate. What is a good option for them to consider using?

    A. Wiki pages

    B. Intranet sites

    C. Collaboration tools

    D. Scrum retrospective

57. The team is working on a cutover to a new badge access system for an agency with a security component. During the planned outage, a problem is discovered that cannot be solved immediately. What should the project team do?

    A. Continue to implement the change.

    B. Implement the regression plan and reverse the changes.

    C. Evaluate the impact and justification of an extended outage.

    D. Identify and document the change.

58. A software development team is sharing their knowledge and progress by using a web browser to edit a common page. Which knowledge management tool are they using?

    A. Intranet sites

    B. Collaboration tools

    C. Internet sites

    D. Wiki knowledge base

59. A project team is working on building a new office complex in a remote area. The on-site office trailer has a telephone line, but internet connectivity is limited and mobile phone coverage is spotty. What would be the best method for team members at this location to participate in a status meeting?

    A. Email

    B. Video conferencing

    C. Voice conferencing

    D. Whiteboard

60. What type of project management tool is depicted here?

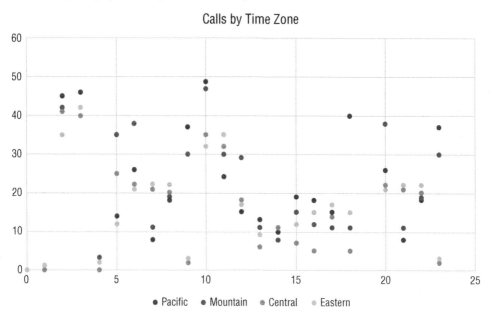

Calls by Time Zone

   A. Process diagram

   B. Histogram

   C. Pareto chart

   D. Scatter chart

61. Which of the following is a simple time management technique where a fixed maximum amount of time for an activity is set in advance, and then the activity is completed within that timeframe?

   A. Sprint planning

   B. Parametric estimating

   C. Timeboxing

   D. Scrum

62. WigitCom is about to launch a new software product to the market. They are concerned about a new regulation that is being considered in a large state that may cause modification to the design, specifically around data privacy. It was suggested during a steering committee meeting that this issue get revisited after the final vote on the new law. What is this an example of?

   A. Action item

   B. Follow-up

   C. Task setting

   D. Refinement

63. *We are here for you!* has done work for DewDrops in the past and has begun to advertise and market that DewDrops is a current project client without DewDrops's consent. What ESG factor might cause DewDrops to issue a cease-and-desist letter to *We are here for you!* to correct this behavior?

   A. Project impact to company brand value

   B. Awareness of applicable regulations and standards

   C. Awareness of company vision, mission statements, and values

   D. Project impact to the local and global environment

64. Time, budget, and scope are considered the triple constraints to tactical project management. What are the three factors that influence sustainable project management from a strategic context?

   A. Environmentally sound, viral social media, economically viable

   B. Economically viable, on time, on budget

   C. Environmentally sound, socially responsible, economically viable

   D. Economically viable, environmentally should, return on investment

65. Which form of quality assurance for a software release would include team members following a checklist and performing actions such as data entry, validating functionality, and viewing reports?

   A. Automated testing

   B. Turing testing

   C. User acceptance testing

   D. Manual testing

66. Which type of change management focuses on service transitions such as moving from one vendor to another or getting all of your departments on board with a new financial system?

   A. Organizational change management

   B. On-premises change control

   C. Cloud change control

   D. Shift change management

67. A company has a large datacenter with nearly a thousand servers and a host of enterprise applications. To align with the company's mission statement, there is a project to move away high datacenter costs while keeping operational consistency. Which cloud environment should be pursued?

   A. IaaS

   B. PaaS

   C. XaaS

   D. SaaS

68. What best describes a request for information?

   A. A meeting with prospective vendors prior to completing a proposal

   B. Procurement method to obtain more information about goods and services

   C. Procurement document that details the goods and services to be procured from outside the organization

   D. Procurement method to invite bids, review, select, and purchase goods or services

69. What is fast tracking a project?

   A. Removing critical path activities that are unnecessary

   B. Moving later deliverables to earlier phases to appease stakeholders

   C. Performing two tasks in parallel that were previously scheduled to start sequentially

   D. Looking at cost and schedule trade-offs, like adding more resources

70. What indicates an individual's or organization's comfort level with risky situations or decisions?

   A. Risk register

   B. Risk tolerance

   C. Risk avoider

   D. Risk taker

71. An adaptive method would be preferable to a more rigid project management style in which situation?

    A. In a mature organization with defined processes

    B. When dealing with a rapidly changing environment

    C. When the scope can be easily and thoroughly defined

    D. Where small incremental improvements offer no value to stakeholders

72. What does a CCB do to support the project?

    A. Helps vet and manage changes to the scope

    B. Provides an accounting structure for tasks

    C. Sets the standards and templates for the project

    D. Sets the costs of quality for the project

73. A project schedule serves what function?

    A. Determines the project cost accounting codes

    B. Creates a deliverable-based decomposition of the project

    C. Lists the actions that should be resolved to fulfill deliverables

    D. Determines start and finish dates for project activities

74. When does an item move from the risk register to the issue log?

    A. As soon as the risk as identified.

    B. When the risk is triggered.

    C. It never ends up on the issue log.

    D. During the creation of the project plan.

75. The project sponsor receives a phone call letting her know that a major task has been completed successfully and that there were no safety issues to report. Which role has the project sponsor been assigned on this task?

    A. Responsible

    B. Accountable

    C. Consulted

    D. Informed

76. Which component of the project charter describes the characteristics of the product produced by the project?

    A. Project objectives

    B. Business case

    C. Deliverables

    D. Quality plan

77. The list of items that need to be monitored and/or escalated to minimize the impact on the project team is called which one of the following?

A. Scatter chart

B. Histogram

C. Pareto chart

D. Run chart

78. All of the following are advantages of using a PaaS EXCEPT:

A. Reduced complexity

B. Self-ramping up or down of infrastructure

C. Reduced control

D. Easier maintenance and enhancement of applications

79. Lashika is the project manager assigned to a project, and she is creating a graphic that shows the project leads for each subject area along with the resources assigned to them. What type of document is Lashika creating?

A. Communication plan

B. Pareto diagram

C. Organizational chart

D. Ishikawa diagram

80. Which tool provides a visual representation of all of the steps required in a process?

A. Process diagram

B. Business process repository

C. Gantt chart

D. Program Evaluation Review Technique (PERT) chart

81. What type of project management tool is depicted here?

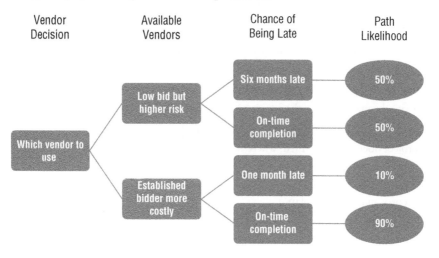

A. Process diagram

B. Histogram

C. Decision tree

D. Scatter chart

82. The sprint planning meeting is used to:

A. Get a head start on the work needed for the project.

B. Prepare the project charter and kickoff meeting.

C. Set a realistic backlog of items completed during this iteration.

D. Set the communication and quality plans for the project.

83. A building project requires the following steps: construction, purchasing the build site, blue-printing, and inspection. Construction has what relationship to purchasing the build site?

A. It is a successor task.

B. It is a mandatory task.

C. It is a predecessor task.

D. It is a discretionary task.

84. Danielle has been doing an analysis to compare any correlation between an independent and a dependent variable. She has created a regression line chart to forecast the changes. What has Danielle created?

A. Process diagram

B. Scatter chart

C. Pareto chart

D. Histogram

85. DewDrops is building a revolutionary new video streaming platform. A functional manager from the administrative division assigned one of his team members to the project and is curious about the project. They ask this team member about details of the project. To conform to the project's access requirements, what should the project team member say to the manager?

A. Share all of the project details with the functional manager.

B. Let the manager review all existing project artifacts.

C. Inform the manager they are unable to share details about the project.

D. Invite the manager into the secure project space and let them look around.

86. As a project manager, which plan would you establish to guide the creation of project files, how project documents are filed, where to archive project records, and when to destroy these documents?

A. Project plan

B. Records management plan

C. Project charter

D. Data capacity plan

87. Pete has been asked to determine the root cause of communication problems on a global project. Which tool should he use to help him with this analysis?

A. Pareto chart

B. Run chart

C. Wiki pages

D. Fishbone diagram

88. The list of items that need to be monitored and/or escalated to minimize the impact on the project team is called what?

A. Issue log

B. Action items

C. Risk register

D. Budget report

89. All of the following are characteristic of an Agile project management approach EXCEPT:

A. Each release is tested against the customers' needs.

B. It uses a flexible approach to requirements.

C. It follows a strict adherence to a change control process.

D. Team members work in short bursts, or sprints.

90. The project team has been asked to identify what will happen if a given IT change has a problem or causes a non-repairable outage. What has the project team been asked to create?

A. Rollback plan

B. Risk assessment

C. Automate tests

D. Validation checks

91. The customer site for a project is located in a rural part of a country, while the project headquarters is located in a metropolitan hub. What would be the biggest impediment to face-to-face communication?

A. Times zones

B. Geographical factors

C. Intraorganizational differences

D. Technological factors

92. About how much time should a project manager spend communicating?

A. Up to 90%

B. Up to 75%

C. Up to 50%

D. Up to 27%

93. During the Closing phase of the project, what two activities are conducted?

    A. Accept project deliverables and perform quality assurance.

    B. Finalize project work and close all vendor contracts.

    C. Manage stakeholder expectations and close all vendor contracts.

    D. Finalize project work and perform quality assurance.

94. What type of project management tool is depicted here?

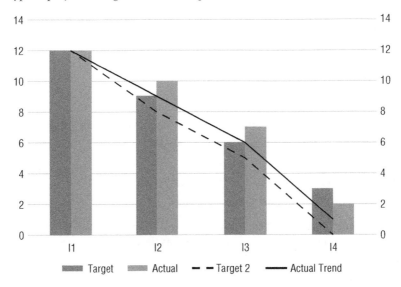

    A. Project network diagram

    B. Gantt chart

    C. Project backlog

    D. Project burndown chart

95. The preliminary scope statement should be included in which project document?

    A. Communication plan

    B. Project schedule

    C. Project charter

    D. Lessons learned

Appendix

Answers to Review Questions

Chapter 1: Project Management Concepts (Domain 1.0)

1.   B, E. A project can be summarized as having the following properties: It is temporary in nature, it creates a unique product or service, it has a definite start and finish, it contains a reason/purpose, and it may be part of a program or portfolio. A group of related tasks is not necessarily a project but could be a to-do list of any kind. Operational activities are activities that take place after a project has been completed. Reworking an existing project is not creating a new product or service, and it doesn't meet the properties of a project.

2.   A. Providing governance on the project is the responsibility of a project management office, not the project team. The project team is responsible for contributing expertise to the project, contributing deliverables according to the schedule, estimating task duration, and estimating costs and dependencies.

3.   A. The project manager is attempting to make the conflict appear less important than it really is by implying they would have to work this weekend if they didn't stay tonight. No attempt is made to see if there was another solution, like coming in early tomorrow.

4.   C. This is where most of the expenditures on materials and people time will occur.

5.   C. This is where both parties give up something to help reach a workable solution. The commitment to meeting the deadline and the agreement to allow for time off but adjusting the timing are the key decision points.

6.   D. The Communication plan is where all the elements of the who, what, when, where, and why of communication needs are documented.

7.   B. A portfolio is a collection of programs, subportfolios, and projects that support strategic business goals or objectives.

8.   C. A program is a group of related projects that are managed together, with coordinated processes and techniques. Option A would be a stakeholder register, which might be a deliverable of the stakeholder analysis. Option B would be the software that is used as a part of the project management information system (PMIS). Option D is the definition of a portfolio. Make sure you know the difference between a project and a portfolio.

9.   D. What did I accomplish yesterday?, What will I do today?, and What obstacles are preventing progress? are the three questions asked during a stand-up, or Scrum, meeting.

10.   B. All authority in this example rests with the functional manager who is pulling Marcus back to work on other assignments rather than the project. In a functional organizational structure, resources typically report to a functional manager and the project manager has limited or no authority.

11.   A. In this scenario, working with the team member to get the desired result is the best course of action. Removing or replacing a team member is not always an option, and thus is incorrect.

12. B. Sprint planning is a tactical activity that is left to the project manager and project team. Steering committees operate in the strategic realm of the project by setting strategies and goals, advocating for different initiatives and projects, and offering their expert opinion to the project team to help guide project success.

13. A. As a general rule, agendas should be published at least 24 hours in advance. Remember that the project communication plan or company norms could set this target to a different threshold.

14. C. In stand-up meetings, participants typically stand, relying on the discomfort of physically standing to help keep the length of the meeting short. Often, stand-ups are held daily and last for approximately 15 minutes.

15. D. Brainstorming is a great way to generate ideas quickly where a group discussion can produce ideas or solve problems.

16. D. In a functional organization, the authority resides with the functional manager, not the project manager.

17. C. The purchasing of the build site must occur before the construction activity begins.

18. C. The Gantt chart is a type of bar chart that shows task duration and dependencies.

19. A. The best answer is communication security. By utilizing the established communication tools, the organizations can enable encryption and multifactor authentication to prevent unauthorized release of information. Communication integrity deals with ensuring information is protected by policies and rules proving the information to be correct and true. Communication archiving refers to making sure records are stored correctly and are deleting according to an established schedule. (Author's note: There is an argument to be made that A, B, and C are all correct as this kind of company or project policy can have positive effects on communication security, integrity, and archiving. So, if you answered one of those options, give yourself a pass).

20. C. An external dependency is where an entity or condition outside of the project drives the scheduling for that task.

21. C. The best answer is communication archiving. Likely established in the project plan or communication plan, the retention of the emails, instant messages, text messages, and other forms of communication should be established. It is likely the company has a retention schedule already established that would just need to be followed.

22. D. A => C => D => E. The critical path has task A (2 days), task C (2 days), task D (2 days), and task E (3 days) for a total of 9 days.

23. D. There is no heroic effort that occurs during an Agile sprint, which is a short burst of activity on a project focusing on a few tasks and working them to a completed state.

24. B. In this scenario, all of the power and authority are present with the project manager and not the functional manager, which is indicative of a strong-matrix organizational structure.

25. B. Status meetings are informative meetings. Informative meetings can also include demonstrations, presentations, and stand-ups.

26. C. Instant messaging is less disruptive than an impromptu meeting where Bridget's concentration might be disrupted beyond just answering the question. Instant messaging allows Bridget to respond and return to her work and is also more expedient than email.

27. B, D. Giving the project team a phone call or sending text messages are preferable because they immediately alert the receiver that you are attempting to communicate with them.

28. B. Well-formed, written communication such as email is a great format to share complex information because the receiver can digest and process the information, refer back to it, and send follow-up questions as needed.

29. A, C. The two most correct answers are that the project has a definitive start and end date and that it creates a unique product or service. Projects are considered a success when the goals they set out to accomplish are fulfilled and the stakeholders are satisfied with the results.

30. B. A portfolio is a collection of programs, subportfolios, and projects that support strategic business goals or objectives. Programs and projects within a portfolio are not necessarily related to one another in a direct way, and projects may independently exist within the portfolio.

31. C. Working on three different continents would mean one work team is likely off-duty/sleeping at any given time during the day. As such, the best method would be email so that the off-duty team would get the information when they begin their shift, while the other teams would get the information instantly.

32. C. This is where the project team members and stakeholders are introduced and the goals for the project are outlined.

33. B. Leadership, time management, team building, and listening are soft skills that are important for a project manager. Critical path diagrams are an artifact, and the creation of them would be considered a hard skill, so options A and C are incorrect. Following and independence are soft skills, but not typically associated with project management, so D is also incorrect.

34. D. The kickoff meeting typically happens at the beginning of the Executing phase and is a primary means of introducing the project team to each other.

35. D. A project schedule determines start and finish dates for project activities and will also have activity durations and order of precedence.

36. C. The Closure meeting will give the status of all activities, allow the project team to turn in any property like ID badges/key cards, and bring that phase of the work effort to an end.

37. A. The project manager is not the critical role in this answer. Unless it falls into one of the other three phases, this does not hold true.

38. B. Where there is a significant change in the situation, interrupting the meeting and letting the CEO know with an in-person meeting would be the best choice due to the urgency.

39. C. Determining activity sequence is an important part of project management, and you will probably see a question like this on the test. For this question, the correct order is: gather bread, peanut butter, and jelly; get a knife; place bread on a plate; spread peanut butter on one slice of bread; spread jelly on the other slice of bread; put both slices of bread together; serve.

40. A, D, F. Some of the important steps required in the development of the critical path include determining the tasks, determining task start/finish dates, determining task durations and milestones, putting them in sequential order by identifying predecessors and setting dependencies, and identifying the critical path. Scheduling activities also address the allocation of resources, setting of baselines, and quality/governance gates.

41. D. For non-urgent information, regular scheduled meetings would the ideal format to share this type of information. On the standing agenda, including a portion to discuss routine updates is a good practice.

42. B. To help control costs and to minimize disruption to project work, a virtual meeting would be the ideal choice to handle routine meetings.

43. A. For routine messages, or for very complex messages, email is the best method as it allows the project team to consume the information when they choose and minimizes the disruption of project work.

44. A, C, E. Using team-building activities, using recognition and rewards, and setting the ground rules are some of the tools that you can use to develop an effective project team.

45. D. This is the ideal state teams are shooting for, where the team is productive and effective. Not all teams are able to achieve this stage of development.

46. D. The project manager should refer to the communication plan and inform appropriate stakeholders that the change has been implemented.

47. A. When requirements are changing, an Agile approach allows an organization to be readily able to adapt to the environment.

48. D. The project manager has the authority to task team resources and conduct performance evaluations, making this a projectized organizational structure.

49. B. The user story helps focus on how the product is going to be used to help shape how it is designed.

50. D. More established organizations with mature processes and tenured staff would most likely have a functional organizational structure centered around specialties.

51. C. Conduct an impact assessment. When a change is being considered, it is important to understand how the proposed change will impact work on the project.

52. B. A backlog is the artifact that is used within an Agile methodology to keep track of all the elements that need to be included in a project but that may not be a part of this sprint.

53. A. By setting fixed blocks of time for the team to reach a decision or direction in the meeting, Thaala is engaging in timeboxing.

54. D. Milestones are major events in a project used to measure progress.

55. B. Joint application reviews are a collaboration that helps ensure the project is meeting its expected goals.

56. A. Facilitators are primarily impartial stewards of the meeting, helping to navigate conflict, address lack of participation, and keep the meeting on time.

57. D. In the Agile methodology, tasks can get recorded and identified in a backlog, but it is not a dependency important to task sequencing.

58. B. Remember that a project is temporary in nature, and at times adding too much to the scope changes the temporary aspect of the endeavor. Adding unapproved scope to the project is considered to be scope creep.

59. D. An agenda outlines the items that will be discussed as well as major decisions that need to be made, and can include materials or topics attendees can read or prepare for in advance to make the meeting more productive.

60. D. This is a meeting dedicated to creating a deliverable, engaging in intensive discussion and/or activity on a particular subject.

61. B. The framing of the walls must begin before the installation of the network cabling can begin, leading to a start-to-start relationship. The wiring must be in place before dry walling and insulation can begin.

62. B. The critical path has A (2 days), C (2 days), D (2 days), and E (3 days), for a total of 9 days.

63. C. The resource calendar will also let you know the dates resources are active or idle.

64. A. Changes to the scope while a project is still active, even if it is modifying the end product, is an example of scope creep and a change to the project.

65. D. Action items are specific tasks or activities that have been agreed on that need to be completed to help with project success. Action items should have an owner and a due date, and this ownership should be confirmed with the owner in the meeting, or immediately afterward if an action item was assigned to someone not in the meeting.

66. D. While Nyssa is assigned to the project, it is clear that the power rests with the functional manager, undercutting the project manager's authority.

67. A. Stakeholder expectations having been met is the most critical factor involved when determining whether a project is a success.

68. B. A deliverable can be any tangible or intangible product or service produced as the result of a project.

69. D. The project sponsors' responsibilities include helping define and develop the high-level requirements for the project, functioning as the approval authority, removing roadblocks, marketing the project across the organization, controlling the direction of the project, and defining the business case for the project.

70. A. It is a successor task. Construction would come after the blueprinting is approved.

71. A. All projects are constrained by what is typically referred to as the "triple constraints," which include time, budget, and scope as they pertain to quality. Typically, you can manipulate two elements and will have to live with how they constrain the third element. You can have cheap, feature-rich, or fast: pick two of three.

72. C. The charter gives formal authorization to begin and to commit resources to the project. Accordingly, project team selection and procurement can begin at this point.

73. B. The scribe captures critical discussion points, action items, and follow-ups to document these items for future reference.

74. C. The WBS is a project artifact, not resources in the resource management context.

75. A. This technique does not change the critical path or project completion date but balances specific schedule dates when there is concern about resource availability.

76. C. Overwhelming individuals on a project can lead to burnout, project turnover, or task slowdowns, which all will negatively impact a project.

77. A. Fast tracking is a schedule compression technique that involves completing two tasks in parallel that were previously scheduled to start sequentially. This can come with a higher cost and with increased risk, but it will help the project be completed faster. Option D might be tempting but remember there are never unnecessary tasks on the critical path and therefore the answer cannot be correct.

78. A. The critical path has zero float or slack time and is the shortest amount of time that a project can be completed.

79. B. There are not enough resources for the task, leading to overallocation. An individual's workload becomes more severe, leading to longer hours and a higher potential for burnout.

80. E. Adjourning refers to the dissolving of the team when work has been completed.

81. A. Kayla should document the request in the change control log. The log can help keep track of the request, decisions that are made on the request, and the status of implementing the request.

82. C. Since the project was already completed, the asset management platform is considered in production. Upgrades, enhancements, and expansions would all be considered changes to the product.

83. A. A more established organization that is working with a fixed budget or other constraints would be best suited for more traditional project management approaches.

84. C. This may be an activity during either the Planning or Execution phase, depending on the nature of the industry the project is in.

85. A. The change control board (CCB), in conjunction with the change control process, will approve or reject changes to the scope of the project.

86. A. While not listed, this function could also be done by a program or portfolio within those units only. The PMO performs this function across the organization.

87. B. Consult the communication plan and inform the appropriate stakeholders. It is crucial that approved changes to the project goals and scope are communicated to appropriate stakeholders. The communication plan should establish who needs to be notified when this event is triggered.

88. D. The change control board helps govern changes made to the scope of a project and help control against scope creep.

89. D. Whenever a team member joins the project, or at the beginning of the project, when multiple people join, is when expectations should be set, including roles, due dates, norms, and team interactions.

90. B. The work breakdown structure is characteristic of a more deliberate, up-front requirements gathering, when many of the requirements are identified early in the project.

91. D. Joint application development sessions help ensure accuracy between the project scope definition and delivery through continuous interactions between the project team and stakeholders. Since the project is still working development, this is not an application review session yet. Note that it is possible for joint application development sessions to use brainstorming or even stand-ups as they work through the project.

92. B. This is the model that was developed by Dr. Bruce Tuckman, and it is known as the stages of team development.

93. A. Daily stand-up meetings are typically used with an Agile methodology. As such, a daily stand-up meeting would not have a governance body in attendance and would normally be attended by project team members.

94. D. The PMO helps keep roles like the project manager and project team accountable from a more strategic standpoint.

95. C. The target audience is the most correct answer as the meeting has been called specifically to demonstrate the project prototype to the community the product is being designed and built for. It is possible that some of these members may be shareholders in the company and might also serve on the steering committee. The purpose of the meeting and size of the invitees makes the target audience the best answer.

96. B, C, E. The three types of organizational structures are functional, matrix, and projectized. In a functional organization, decisions and control are driven by the specialized function (IT, accounting, HR, and so on). With a projectized organization, the project manager has decision-making authority and control over resources. A matrixed organization is a blended model between functional and projectized.

97. C. The project steering committee is the senior decision-making body within the project's governance structure.

98. A. Follow-ups check on the status of any new developments in the project environment that may have an impact on the scope, cost, or success of the project. These developments can be either internal or external.

99. B, D, E. The PMO provides guidance to project managers and helps present a consistent, reliable approach to managing projects across the organization. Responsibilities include providing governance for projects, maintaining standard documentation and templates, and establishing key performance indicators and parameters.

100. A. The "champion" role of the sponsor is really important both initially and as the project commences in order to keep the energy and focus of the whole organization committed to its success.

101. C. A program is a group of related projects that are managed together with coordinated processes and techniques. Make sure that you know the difference between a project and a portfolio.

102. C. Minutes help capture when and why key decisions are made, which is helpful to refer back to later in the project. Capturing decisions and action items can be critical for project success.

103. B. When presenting findings or product functionality, a demonstration allows for finished, if not final, artifacts to be shared with interested stakeholders.

104. B. Focus groups involve interviewing a small number of people with common traits, experiences, or demographic similarities.

105. C. This helps create the differing levels of decomposition.

106. B. Finish-to-start is the type of logical relationship most frequently used within precedence diagramming methods.

107. B. This stage is a process of establishing who is the most influential, and there is jostling for position.

108. C. Timeboxing can be either hard or soft. Hard timeboxing dictates that the task or activity must stop when your time is up, regardless of the status. Soft timeboxing is more flexible but gives the expected time a task should stop. This can be useful in complex tasks or activities.

109. D. Hard timeboxing can be helpful in having a meeting move on against the desire for perfection or creative churning over possibilities.

110. B. This can be recorded through the use of user stories or short descriptions of the functionality.

111. C. Alternative analysis gauges the different options that might be considered to accomplish the assignment with the resources that are available.

112. B. The project management office's responsibilities include setting standards and practices for an organization, providing tools such as previous project documentation, and delivering standardized documentation and templates.

113. B. Crashing involves looking at cost and schedule trade-offs such as adding more resources.

114. A. Fast tracking involves performing two tasks in parallel that were previously scheduled to start sequentially.

115. D. Pareto diagramming, or producing a Pareto chart or diagram, is a tool used to focus attention on the most critical issues. It is not used to estimate activity duration.

116. B. If you only have a structural engineer or senior programmer's time for a small window, the schedule is engineered to make use of that resource window.

117. B. Contingency reserves are calculated for known risks and are paired with documented mitigation response plans.

118. E. The kickoff meeting typically occurs when the charter is signed. The CompTIA Project+ objectives identify it as occurring in the Initiation phase but it is common to see it at the beginning of the Execution phase as well.

119. D. Project managers can spend up to 90 percent of their time communicating with the stakeholders and the project team.

120. B. Vested interest, providing input and requirements, project steering, and expertise are examples of stakeholder responsibilities. Stakeholder expectation setting and engagement are key elements to project success.

121. A. In this case, Jenny is considered a project stakeholder. Remember that a team member is also a form of a stakeholder.

122. D. Contingency reserves are set aside to deal with risk consequences, and management reserves cover future situations that can't be predicted.

123. D. Closure meetings are where the final handoff of the project is conducted, including closing out any contracts and sometime conducting lessons learned. Note: Lessons learned meetings can be separate from closure meetings.

124. A. The confidential nature of HR issues, plus the need to handle a situation delicately, would invite a face-to-face conversation. A good rule is praise publicly and criticize privately.

125. D. The project manager is responsible for all project artifacts produced during the course of doing project work.

126. B. Characteristics of deliverables that must be met are known as requirements. Distance traveled on a project is good information for expense tracking and reporting. Checkpoints are gate checks on a project. Major events to track progress are milestones.

127. A. Define activities, sequence activities, estimate resources, and estimate duration are all key activities that must occur in order to develop the project schedule.

128. A, B, E. Of the provided list, time zones, cultural differences, and language barriers would be factors that influence the project the most. Level of report detail, criticality factors, and technological factors would not be unique to an international project.

129. A. Language barriers can present huge obstacles in communication, sometimes even when the same language is being spoken in different dialects.

130. C. The sprint planning meeting establishes what can realistically be accomplished during the sprint.

131. D. The more involved nature of instructions would allow written, detail communication to reach the trailer. Other electronic forms of communication would not be reliable in this situation.

132. B. There is a safety aspect to this change where Phil or Fernando could come to bodily harm. Therefore, getting them on the phone and ensuring they know the delay is happening is the superior method of communication.

133. A. Due to the geographic separation of the project team, the most efficient way to get the communication to the entire project team would be email. Attempting to schedule a virtual meeting would be troublesome due to the different time zones of the project team.

134. A, D, E. When dealing with any team on different continents, or even on opposite sides of the same continent, time zones will impact communication. The project team also needs to account for language barriers and cultural differences when planning communication.

135. B, D, G. The basic communication model is sender-message-receiver.

136. C. Organizational changes would come from the functional side of the organization, not the project side.

137. A. A change request log can be a simple spreadsheet that has the ID number, date, name, and description of the change.

138. C. A template allows for a standard submittal in the change control process that includes all the information needed on the purpose and impact of the change.

139. B. In a projectized organizational structure, the project manager has the most authority, and resources report directly to the project manager.

140. B. Rather than start-to-deferred, the missing relationship is start-to-start, meaning that one task must wait for a different task to start for work to begin.

141. C. A traditional waterfall methodology can be progressively iterative, similar to the iterative approach used with the Agile methodology.

142. C. Weather conditions are examples of dependencies that are external to the organization but that need to happen for a project or activity to proceed.

143. A. Gate checks would be more representative of a waterfall approach to project management whereas Agile uses a more adaptive approach to requirements and execution.

144. A. Due to the temporary nature of projects, resources are not permanently assigned to any one project or function.

145. C. They are costly because individuals are being paid to sit around. This often happens in projectized organizations.

146. D. The other options would not take advantage of the flexible, easily changing environment of the project.

147. C. A rapidly changing environment like that found in a start-up would be the ideal candidate to use an adaptive approach.

148. D. Early in the process is when individuals will be shy, reserved, and treat each other formally.

149. C. Authority and power are shared between the functional manager and the project manager.

150. B, D. This is an example of ongoing operations that are not temporary in nature and that do not have an end date. Additionally, the work effort does not produce a unique product.

151. D. With values born out of Lean principles, this is an iterative, incremental approach to managing the activities on a project.

152. C. There are times where the best course of action for the entire project team is to move on without the disgruntled team member.

153. A, D. An organized effort to fulfill a purpose and has a specific end date are the most correct answers.

154. D. The project team contributes their expertise to the project and gives their estimates for task duration, cost estimates, and dependencies.

155. C. Also known as a person-hour estimate, this is used in the creation of cost estimates.

156. C. The iterative nature of an Agile approach, which emphasizes constant feedback and interactions, would be the best approach in this situation.

157. D. Tiffany is overallocated for the project assignments because there is more work to do than time to work on it.

158. D. Reporting to both the project manager and the functional manager means that Amy is a shared resource.

159. A. A discretionary dependency is defined by the project management team, and they are normally process-or procedure-driven.

160. C. Effective team building can help create efficient and effective groups focused on getting the work done, oftentimes enjoying the work immensely.

161. B. Commonly a bar chart, this depicts the burndown chart that maps the progress made with each iteration as the project approaches completion.

162. D. The communication guide sets who needs to be informed of what types of project changes and will direct the method that communication should be delivered.

163. D. This is a form of a lessons-learned meeting used in the Agile methodology to help improve future sprints.

164. B, D, E. This meets the requirements for a project because WigitCom is creating a new product or service, and the effort is temporary in nature. It would also be a part of a program, or related projects, that share resources. In this case, the security team's resources would be shared with this project.

165. B. With an Agile approach, the result is that requirements can be adjusted as results are developed during the iterations.

166. B. Work slippages can impact the timeline, especially when that work is along the critical path.

167. A. Highly detailed information that requires time to digest is best suited in writing. Hence, the work lead tailors the message delivery based on the content of the message itself.

168. B. The burndown chart is a visual representation and measurement tool showing the completed work against a time interval to forecast project completion.

169. C. The CCB, not the SME, would be the role whose responsibility is to approve or reject the change.

170. D. A subject matter expert, with input from the project manager, should evaluate the impact of the change to determine if the change should be attempted, the impact to the triple constraints, the benefits, and so forth.

171. C. After the team has formed and stormed, this is where familiarity with one another helps to settle things down and individuals begin to deal with project problems instead of people problems.

172. C. Anyone working on or associated with the project can submit a change, which is then reviewed by a change control board or other process. As such, an executive sponsor is not needed for a change request.

173. A. Change requests should always be captured in writing to document the request and the response to the change, and to keep a paper trail to show why the project's timeline, budget, or scope might have changed.

174. B. The project manager has full authority and controls time and tasks. As a project manager, you need to be wary of having low-quality resources assigned to a project, because sometimes functional managers will try to give you their lowest-caliber individuals. Project managers need to be given full authority to ensure the right skill sets are allocated.

175. B. An example of a mandatory dependency might be pouring the concrete foundation and letting it cure prior to framing a building.

176. B. Key stakeholders, project staff, sponsors, or governance/CCB members could all be the ones who submit a change request for consideration.

177. D. After approval, the next step would be to implement the change. At that point conducting a quality check and updating documentation can occur.

178. D. The personnel on the project have switched so this is a resource change.

179. B, D. The scope and funding were the common elements that were changed in this example.

180. D. The communication plan sets out who needs to be informed of what types of project changes, and it will direct which method of communication should be used.

181. D. Project sponsors should use their influence to help remove obstacles like nonparticipation on the project.

182. B. The requirements to be met for the project have changed but have not fundamentally added to the scope of the project.

183. A, D. Adding more work to the project that is not part of the baseline is a scoping change. The added time would be a timeline change.

184. C. Fast tracking is a schedule compression technique where activities or phases decided to be done in sequence are instead done concurrently.

185. A. Organizations where the project managers have authority of personnel and other resources is a projectized organizational structure.

186. B. This helps depict the differing levels of decomposition depending on the expected date of work.

187. A. The project scheduler is responsible for communicating the timeline and any changes.

188. C. A Scrum master is responsible for the direction of the work and the team and helps ensure quality and communication throughout a project. This position needs skills in Agile and Scrum and should have good experience in both before becoming a Scrum master.

189. B. Budget is the cost constraint, and the team will not be able to work unlimited hours on the project without it affecting budget, scope, or quality.

190. A. Product managers are advocates for customers who ensure products add value to the lives of consumers.

191. A. This is not an element of the scope baseline but a management skill used to help get teams unstuck.

192. C. The iterative nature of an Agile approach, which emphasizes continuous feedback and interactions, would be the best approach in this situation.

193. A. This is indicative of a more traditional waterfall approach where the scope is controlled with a more rigid change process.

194. C. The waterfall model of project management moves downward through each phase and the phases of development do not overlap. Rapid feedback, embracing change, and assuming simplicity are all principles of extreme programming, which is intended to improve software quality and responsiveness to changing customer wants and needs.

195. C. The sprint planning meeting sets what can realistically be accomplished during the sprint.

196. A, B, E. Agile is characterized by self-organized and self-directed teams, sprint planning, and continuous requirements gathering. Projects using a waterfall technique would have characteristics of formally organized teams and saving feedback for lessons learned meetings at the end of the project.

197. C. DevOps is used to allow formerly siloed roles, including development and IT operations to coordinate and collaborate to produce better and more reliable products.

198. A, B, D, E. Option C is not correct because that is the responsibility of the project management office, and option F is the function of the project manager.

199. C. The five main components are Architecture, Integration, Governance, Funding, and Roles. These components help provide a structured and methodical approach to project alignment and completion. Each of these components can be considered on three different levels: team, program, and portfolio.

200. D. PRojects IN Controlled Environments (PRINCE2) provides a methodology to perform and complete the project. The Project Management Professional (PMP) provides techniques, tools, and frameworks to perform and complete the project. The Software Development Life Cycle can utilize several methodologies, including Agile and Waterfall to lay out a plan for getting a software development project correct the first time. Extreme programming is also a software development methodology focusing on producing high-quality software in an environment of evolving and changing requirements. Since B and C are software development methodologies, D is the correct answer for project methodologies.

201. B. Kanban is a lean software development methodology and organizes work using a Kanban board. A Kanban board organizes work in columns, which identify the states every work item passes through from left to right. Kanban manages the work in progress (WIP) and optimizes the flow of work items through the project.

202. D. DevSecOps stands for development, security, and operations. DevSecOps commits to application and infrastructure security consideration from the beginning of a product's creation. Kanban, Agile, and DevOps are good answers, but the financial nature of the start-up implies that extra security protocols are needed, making DevSecOps the best answer.

203. B. Projects utilizing the Waterfall methodology have self-contained stages like requirements, design, and implementation and generally one stage doesn't begin until the previous stage is completed.

204. C. Low-quality resources do not have the required experience, are not accomplished with the skill set, have poor passion for the project, or carry a bad attitude to work. Low-quality resources can sometimes be replaced, but they will always need to be managed in some form.

205. A. Requirements, Design, Implementation, Verification/Testing, Deployment/Maintenance are the common stages in a Waterfall process.

Chapter 2: Project Life Cycle Phases (Domain 2.0)

1. D. The business case can include justification, alternative solutions, and alignment to the strategic plan.

2. D. The communication plan is where all elements of the who, what, when, where, and why of communication needs are documented.

3. D. The high-level scope definition sets the boundaries of what is included and what is specifically excluded. This includes the deliverables, objects, and purpose of the project.

4. C. The Planning life cycle phase sets the details for how the project will be carried out and creates the majority of the project plan artifacts, including the project schedule, risk plan, communications plan, and other pertinent documents.

5. A. The project description explains the attributes of the product, service, or result of the project.

6. D. While stakeholders start out with a lot of influence, it decreases as the project advances.

7. A. Forcing is where one party gets their way, and the other party's interest is not represented.

8. A, D, F. Commonly referred to as the "triple constraints," almost all projects are constrained by time, budget, and scope as they impact quality.

9. B. Similar to the availability of people with certain skills, equipment availability and *scheduling* can also lead to a constraint on a project.

10. B, C. Two of the triple constraints are represented in the options: a predefined budget and a fixed, or mandated, finish date. These constraints will impact the options available in terms of scope, and they should be considered in terms of creating the scope statement.

11. E. Other closing activities would include getting project sign-off, validating all deliverables, closing the contract, removing project team access. and conducting training as a part of the transition/integration plan.

12. B. The project charter is prepared and agreed to in the Initiation phase of a project.

13. C. The kickoff meeting is where the project team members and stakeholders are introduced.

14. B. The high-level risk assessment is an input to the project charter. The risk management process helps determine the approach to addressing risks, and the initial assessment included in the charter gives a broad overview of the risk landscape for the project.

15. B. Constraint reprioritization is a shifting of which constraint is unmovable, thereby setting how the other constraints are manipulated. The old adage "You can have it cheap, right, or fast; you get to pick two of three," illuminates the interplay between the triple constraints.

16. C. Planning includes identifying both what resources are needed and when they will be needed.

17. C. The confronting technique may also be called problem solving and is a superior way to resolve conflict. The correct solution to a problem can reveal itself, and the facts contribute to discovering the solution.

18. A. Developing a team that lasts longer than the project would be a nice outcome, but it is not the purpose of developing a project team.

19. B. WBS components should always happen concurrently with determining major deliverables. Identification of lower-level WBS components occurs after the major deliverables have been identified.

20. B. Compromising is when each party in a conflict gives up something to reach a solution.

21. C. CapEx is important to help companies grow and maintain their business by investing in new property, equipment, products, or technology.

22. A. Capital expenses are used on spending for a benefit that will last longer than a year, and some companies might have a dollar amount that a single item might cost that would also lead to the purchase being qualified as a capital expense. As an example, a company might have a $5,000 limit on the individual cost of an item, meaning the purchase of a $6,000 server would qualify as a CapEx but a $2,000 desktop computer would not.

23. A. Examples of OpEx would include rent, office equipment, supplies, legal fees, travel expenses, and utilities, to name a few.

24. B. Expenses with a business partner for the time and travel expenses would likely be categorized as operational expenses. For the test, it is unlikely there will be complicated scenarios, but know that in the real world there are situations where this example could be categorized

as a capital expense if the work being performed is part of the implementation of a long-term asset, like a datacenter build, enterprise resource planning implementation, or other large-scale purchase.

25. D. This is a capital expense for both the length of time the new building will be used and the dollar amount of the new investment. Note: A business will establish rules where a certain dollar amount threshold (as low as $100) makes a purchase a capital expense.

26. C. Terms of reference in a contract explain the objectives, scope of work, activities, tasks to be performed, and other information such as the structure of a project.

27. A, C, E. Terms of reference help provide context and information need for bidders to provide comprehensive proposals when answering an RFP or other solicitation. It should contain information about your company, the goals of the project, budget (if known), timeframe, and standards to name a few areas that should be included.

28. A. A governance gate is a checkpoint between project phases where approval is obtained to move forward. Usually the project reports to a steering committee to help ensure accountability on the project for time, money, and scope.

29. B. The prequalified vendor list helps buyers prescreen interested parties that sell products and services. This can help create a pool of prospective sellers, which can speed up future procurement processes, establish relationships, and increase the chances a seller will be noticed and selected for future business.

30. A. Becoming a prequalified vendor would allow Jazzmyn's company to pass a first phase of vetting, narrowing the seller pool and thereby increasing the chances her company is selected to perform work.

31. B. A return on investment (ROI) analysis can be used to determine if there will be a positive return on a project's investment.

32. D. ROI can help compare a project investment to other well-understood risks available in financial terms to help determine the profitability of the proposed solution.

33. B. For the exam, you may need to know how to calculate ROI. If the value generated from this equation is less than 1, there is no profit or financial gain from the project. If the value is 0, there is no loss or gain. A value greater than 1 indicates there is a profit or gain from the project.

34. B. A positive number will indicate the project will be profitable. Remember, a number can still be positive and too low to still be considered a "go" project by the organization. The goals and risk threshold of the company will help guide the interpretation of these results. A number greater than 1 in an ROI analysis will indicate the project is profitable.

35. A. A number less than 1, in this case a negative number, will indicate the project will not be profitable. Just because a project will not be profitable does not necessarily mean it should not happen, just as a profitable project is not always assured of being done. ROI analysis is just one of many tools to help determine the viability of a project.

36. B. Managing project records is an important responsibility of a project manager, who must make sure every document or file has the proper design, format, communication, and security and is archived and stored correctly.

37. B, C, D. A records management plan ultimately supports the efficiencies in all other project activities by making sure data is secure, easy to find, and up-to-date and versioned.

38. A. Defining access requirements early in a project can help protect the confidentiality of documents and data as well as protect intellectual property.

39. C. The access requirements for the project are a guiding reference to what information can be shared and to whom. The complexity in this scenario is that the manager is the functional boss for the team member, and the familiarity may make it tempting to share details about the project.

40. C. Communication channels are likely to be established to ensure information governance, confidentiality, and information retrieval.

41. E. Archiving documentation is an activity of the Closing phase.

42. A, D. Developing a transition plan occurs in the Planning phase for when the project completes and is handed over to operations. Implementing organizational change management is an activity that would occur in the Execution phase of the project.

43. A. RACI stands for responsible, accountable, consulted, and informed, and it is a tool to help designate roles and responsibilities on a project.

44. C. The resource calendar will also let you know the dates resources are active or idle.

45. C. The WBS is a project artifact, not a resource in the resource management context.

46. A. The scope of the project has changed but not the schedule, so costs will increase. This is an example of the triple constraints driving what is important in a project.

47. B. The procurement plan document details how the procurement process will be managed.

48. D. Deliverables for the project are produced in the Execution phase.

49. C. The lowest level recorded in the WBS is the work package.

50. B, E. Smoothing and negotiating are conflict resolution techniques.

51. B. Storming is the process of establishing who is the most influential; there is jostling for position.

52. D. The procurement plan document includes the approach to obtaining outside products and services, and the rationale for choosing to in-source or outsource the effort.

53. A. Project sign-off occurs in the Closing phase, not the Initiation phase.

54. B, C, E. The communication plan, transition/release plan, and project schedule are all created in the Planning phase.

55. B. These are characteristics of deliverables that must be met.

56. A. According to the *Project Management Body of Knowledge*, the steps taken to create the project schedule are: define activities, sequence activities, estimate resources, and estimate duration.

57. C. The sprint planning meeting sets what can realistically be accomplished during the sprint and defines units of work.

58. C. The scope statement in the project charter sets the boundaries for the project.

59. B. The bulk of the work on a project is carried out during the Execution life cycle phase, including the creation and verification of deliverables.

60. B. It stands for responsibility, accountable, support, and inform. This is an alternative form of a RACI chart, with "consulted" replaced by identifying who serves in a supporting role.

61. A. The explanations of team members' roles and responsibilities would generally be found in a responsibility assignment matrix instead of the WBS dictionary.

62. D. Scope creep is when the project definition is constantly expanding and control over being able to finish the project lessens.

63. B, C, D. The scope management plan includes the process for creating the scope statement, the definitions of how the deliverables will be verified, and the process for creating, maintaining, and approving the WBS. It also will define the process for controlling scope change requests, including the procedure for requesting changes.

64. B. Short stories about someone using the product or service help to focus on how the product is going to be used and help shape how it is designed.

65. D. Execution phase is when most of the expenditures on materials, people, and time will occur.

66. D. The preliminary scope statement includes, among other things, the project objectives, criteria for project acceptance, requirements and deliverables, and milestones.

67. D. Risk identification of all potential project alternatives would take place later in the project when analyzing and creating a risk management plan. The preliminary risk assessment is a high-level overview and an input to the project charter.

68. E. This is the Closing phase. During the transition from the project to normal operations, training of the new specifications will occur.

69. B. A good practice when identifying project risks is to have a set of questions to ask during the brainstorming step.

70. C. After the team has formed and stormed, norming is when the familiarity with one another helps to settle things down and individuals begin to deal with project problems instead of people problems.

71. B, C, E. The ways to organize the WBS are by subprojects (where the project managers of the subprojects each create a WBS), by project phases, or by major deliverables.

72. C. A project assumption can best be described as factors considered to be true for planning purposes. Assumptions are events, actions, or conditions that are believed to be true.

73. A. It creates a shared understanding of what is included and excluded from the project. On the test, high-level scope definition might also be referred to as a preliminary scope statement.

74. D. During smoothing, no real or permanent solution is achieved. This can be an example of a lose-lose result as neither side gets resolution to their issues.

75. C. The lessons learned session can occur during any phase depending on the project size but is typically included in the Closing phase of the project.

76. B. A project is represented by the individuals on the team, and ensuring the correct skills, level of experience, availability, and interest are all part of the team selection process.

77. A. Remember that resources can be people, materials, or equipment.

78. A. Much like the organizational breakdown structure, the RBS lists the different resources by categories like labor, heavy equipment, materials, and supplies.

79. D. This is especially true when tasks are along the critical path, or when there is no float for that activity.

80. A. Milestones can also represent the completion of major deliverables on a project.

81. B. One of the most common risks to any project can be staff turnover, and when this happens it needs to be communicated to the project team and the project manager.

82. A, C, E. Auditors are a project independent group that wouldn't need to be communicated with during the project. Product end users would only be communicated with when the product was ready. Shareholders are not normally involved in a project when minute details like milestones would be communicated.

83. C. An organizational breakdown structure (OBS) is a hierarchical model setting or it describes the established organizational framework for project planning and resource management.

84. C. The transition or release plan should be created in the Planning phase. Starting this work early allows for the end state of the project to be imagined for the transition to operations.

85. A. Ensuring the operations team has the knowledge and skills to perform the work after a project is completed is a key activity of any project.

86. B. Understanding the steps to actually be performed when a new product or service is launched is critical to confidence in the project deliverables as well as key to organizational change management since people will believe in a well-run rollout more than one that has lots of errors and misses.

87. D. Whenever new resources are brought on a project, it is important to train them on the performance expectations, standards they must follow, and reporting requirements so that the project can run quickly and effectively.

88. B. When beginning a project, assessing what skills and staff are available helps determine gaps that need to be filled for the project to finish on time. Evaluating the vendor pool for cost and availability of staff and services to fill the gap is done in a preliminary procurement needs assessment.

89. C. The requirements for the house outline the success criteria for the project. Other requirements might include when the construction must be completed by, the total cost/budget, passing inspections, and no permit violations. Success criteria will vary from project to project.

90. D. Access requirements will pertain to both physical and electronic access to project materials and work locations. In matters of intellectual property, protecting information from unwanted eyes may be critical to a company's success.

91. B. As the project is developing from an idea to an organized work effort, reviewing existing artifacts such as previous project documentation, feasibility studies, and company documents can help inform the parameters and activities needed for the new project.

92. C. A quality assurance plan helps develop the processes, standards, and validation needed for the creation of a product that meets or exceeds the quality target of a company, project, or product.

93. D. The target audience of a transition plan/release place oftentimes will need to include both internal and external stakeholders, depending on the nature of the project. It is important to assess early who will need to be consulted or informed of changes to the status quo.

94. A. Organizational change management is a critical part of the adoption of a new product, service, or business process. A project needs to account for the concerns, fears, needs, and expectations of stakeholders to counter potential resistance and ensure project adoption. It does not matter that a solution is excellent if the adoption of that solution or change is not also excellent.

95. D. A key part of the Execution phase is the management of vendors supporting the project. The project manager must ensure the expectations spelled out in procurement or project documents are followed and enforced.

96. B. A statement of work formally captures and defines work activities, timeframes, milestones, and deliverables that a vendor must meet in the performance of work for a customer.

97. D. A request for quotation is generally more involved than just a price per item response. It can also be called an invitation for bid.

98. D. The formula used is (Most Likely + Optimistic + Pessimistic) / 3.

99. A, B. To ensure maximum participation by Don on the project, aligning the communication approach to his preferences will help with his engagement.

100. C. The project will no longer be able to keep to the current schedule due to the emergency.

101. D. It is important to attempt to meet stakeholder preferences in terms of communication so that you get an active response and high levels of participation from that stakeholder.

102. B. A resource calendar describes the timeframes in which resources (equipment and team members) are available for project work.

103. B. The 95 percent phenomenon is where the project seems to stagnate or be unable to complete the last step or steps so that the project can be completed.

104. D. Version control helps create the audit trail of changes to documentation, when decisions change, and when the last update to certain documents took place.

105. B. A change control board (CCB) is the body that evaluates and approves changes.

106. A. Service-level agreements spell out the thresholds that must be maintained in providing services, such as response times, defect rates, and other qualities.

107. D. Gate reviews are presented to a governance board or steering committee to validate cost models, schedule, risks, and objective obtainment.

108. C. Since the project team is working on the proposed project schedule, the trigger is project planning. If the situation was adjusting the schedule after the project is already underway, then the answer would be schedule changes.

109. D. During the execution of the activity, the expectations that were set have been changed. This should invite the project manager to reset communications with all stakeholders so they will know when the service will be restored.

110. B. With the identification of new risks, the information should be captured in the risk register and stakeholders need to be informed so that mitigation plans can be developed.

111. B. Communication occurs throughout the project planning process in the creation of plans for risk, quality, communication, schedule, budget, and scope, to name a few.

112. A. Crashing is a technique where resources are added to the project at the lowest possible cost, reflected in the decision to use overtime instead of adding team members.

113. B, D. Sometimes, the urgency or importance of certain communication can overrule a stakeholder's personal preferences because the project team can be at a standstill unless a decision is made.

114. D. The practice of adding initials at the end allows readers to see who has added comments or changes to a document. Modern productivity and collaboration tools can be utilized to help keep track of version control automatically.

115. D. Defects, rework, and their related costs are all related to the quality plan, or a quality change.

116. A. The key to this answer is the approval to move forward. Gate reviews are check-in points where authorization to proceed must be given.

117. B. Ralph's role on the project is that he is accountable for all actions on the project and those of the project team. He has the authority to task, delegate, and ensure the quality from the deliverables.

118. C. A purchase order is an official offer issued to a seller indicating types, quantities, and agreed-upon prices for products and services.

119. B. It is critical to capture any assignments that have been handed out, who owns the assignment, and when the task is due. This helps make sure there is no lapse of completion due to unclear assignments and lack of clarity on what needs to be complete by when.

120. A. A status report is a written document that captures the progress being made on a project, what challenges have materialized, and data on tracking to completing on time and on budget.

121. D. The project charter is the document that authorizes the project to commence.

122. A, B, C. Meeting agendas, communication plans, and project charters would all fall into the category of planning documents at various levels. Project condition information, or status, can all be gleaned in different forms from dashboard information, status reports, and meeting minutes.

123. C. A request for proposal is a formal document that an organization posts to solicit bids from potential suppliers and vendors.

124. A, B. It is possible for someone to be both responsible and accountable. Remember that there can be only one person who is accountable, but multiple people can be responsible. Where an individual gets assigned to multiple participation types will be dependent on the number and quality of staff assigned to the project.

125. E. For the exam, know this activity is identified in the Closing phase. Depending on the size of the project and the culture of the company, it is possible that big milestones or other events might be causes for rewards and celebrations. As former Walt Disney World Vice President of Operations Lee Cockerell says, if you burn the "free fuel" and appreciate, recognize, and encourage everyone, you will have higher morale, productivity, and loyalty.

126. B, C. To help the project team members grow, it is important to give honest feedback in their performance: things they did well and things they could do to make them even better. Additionally, they should be released from their project responsibilities so they can return to their functional jobs. If they are contract resources, they should be freed up to return to their company, which can assign them to future work.

127. A. Following the principle of least privilege, make sure that physical and digital access is reviewed and revoked for any project personnel who no longer require this access. This is vital to help ensure the security of the product or service once in production.

128. C, D. Ensuring the stakeholders have a voice in the project will help to capture lessons learned that can be utilized on future projects. Additionally, checking in with the stakeholders allows for validation and discussion on how successful the project was in satisfying the objectives.

129. D. It is vital to verify deliverables to ensure they meet specifications and satisfy project objectives. Additionally, vendor payments are often tied to deliverables, so before the vendor is paid and let off the hook for performance, make sure the deliverable is what the project expected it to be.

130. D. An activity of the Closing phase, the project manager validates all project expenditures, closes purchase orders and contracts, and finalizes the total spend for the project.

131. C. A Gantt chart is a tool that displays activities and represents visual task duration along with activity precedence.

132. C. To close out the project, it is a good idea to describe the status of the project at completion. This can include known additional risks that might need to be monitored on the new product or service. It might also include future scope that could be taken up in a future project that wasn't completed or developed as an opportunity to pursue.

133. B. A project manager can spend up to 90 percent of their time communicating, and the ability to provide ad hoc status updates can provide timely information to help manage anxieties and expectations.

134. A. Team touchpoints are an effective mechanism to determine the obstacles and momentum of a project. A team touchpoint doesn't have to be a meeting since collaboration tools can allow for asynchronous status updates to be shared with the project manager and the rest of the project team.

135. B. Overall progress reporting helps to keep key stakeholders informed on the project's momentum, spend, and challenges as it works through the work activities.

136. A. External status reporting is targeting stakeholders outside of the project organization such as the organization's board of directors. Other forms of external status reporting could be to a government agency that awarded a grant to an organization to complete certain work.

137. C. Monitoring performance is an activity of managing vendors. Making sure vendor resources are on track for their assignments is as important as it is for internal project resources.

138. C. Risk identification, probability and impact analysis, and risk response would all take place during the Planning phase of the project.

139. D. A gap analysis compares the progress to date versus the expected progress to see if the project is on schedule and on budget.

140. A. A risk can be either positive or negative, which can guide the project's response to dealing with a risk should it occur. Gabriel should be using both the risk plan and the risk register as inputs in the creation of his report.

141. C. A gate review is a formal governance step designed to ensure required processes and tasks have been completed and that the project is worth continuing.

142. A, D, F. The sprint goal sets the single objective and commitment of the developers. The backlog brings transparency to all work that is needed to be performed. The action plan for the sprint is dynamic and changes as the sprint progresses.

143. D. Probability and impact is a prioritization tool and not a category that might be included in a resource breakdown structure.

144. B. RACI stands for Responsible, Accountable, Consulted, and Informed. At a task or activity level, it identifies who needs to do what action based on the individual's role.

145. C. Being consulted means there is a two-way communication that will take place prior to a decision or action being taken for a particular task or decision. This will often vary in a RACI matrix from activity to activity.

146. D. By requiring the successful vendor to carry the insurance, it allows the consequences of negative risk to be transferred to the third party.

147. C. The organizational chart is a graphic representation of how the project is structured and contains the reporting relationships for the project.

148. C. This defines a status report. While this is a specific type of communication, the need for the report and who it will be sent to would be included in the communication plan. The actual report itself is a different project artifact.

149. B. For any given task, there must be one, and only one, position responsible to be accountable. Being accountable in a RACI context means the one person ultimately answerable for the correct and thorough completion of a deliverable or task.

150. D. This plan sets the written communication, meeting schedule, and escalations that should occur to keep the identified stakeholders up-to-date on project events.

151. D. A status report or dashboard would allow the stakeholder to get the information on the current conditions of the project. Other performance measurement tools like a balanced scorecard could help give a glimpse into the project.

152. B. Milestones are major events that are used to measure progress on a project.

153. C. With a RACI matrix, it is easy to get accountable and responsible mixed up and when working through the creation of the matrix. Accountable means they are the ones who must see the task completed, whereas responsible means the people who will do the work. Keep a sharp eye out to make sure you don't get confused.

154. D. The issue log contains a list of issues with IDs, descriptions, and owners of the issue, and it starts to develop as the execution of the project gets underway.

155. C. Governance gate meetings help to ensure accountability and that the project is proceeding according to plan.

156. A. A Scrum introspective is a form of lessons learned done with the Agile methodology at the end of each sprint.

157. D. Meeting minutes are intended to keep track of details of conversations and decisions that were made in the moment. Often during a project, as time passes and with so many other decisions made since a particular decision, it can be difficult to recall what was done and why. Meeting minutes are a tool to help address those challenges.

158. D. This stakeholder is designated to be informed, which is a one-way communication that takes place after an action or decision has already been completed.

159. D. During the kickoff meeting, formal approval for the project might occur in the form of project sign-off.

160. A. The criteria for approval are those conditions that must be in play for the customer to accept the end results of the project.

161. A. Deliverables are specific items that must be produced in order for the project to be complete, and they are usually tangible in nature.

162. C. Assumptions are those things believed to be true for planning purposes. It is important to communicate assumptions with key stakeholders so there are no missed expectations.

163. D. The timeframe, need for a mobile app, and security are all examples of a project's requirements.

164. D. An action items list depicts all the project actions that should be resolved in order to fulfill deliverables.

165. B. For longer and more complex projects, status reports become key to make sure stakeholders are current on the progress, risks, and issues facing the project.

166. B. Constraints are those conditions that restrict or dictate the actions of the project team.

167. D. Activities that monitor the progress of the project and that require corrective actions occur in the Execution phase.

168. D. When someone has a role of informed, it means they are informed after a decision has been made or a result has been achieved.

169. A. Formal approval is normally a requirement for purchase orders to be created, for team members to be assigned to the project, and for work to begin.

170. A, C. The defining characteristics are a definitive start and end date, and creation of a unique product or service.

171. B. Project stakeholders have a vested interest in the project or deliverables and provide inputs and requirements during the project planning. Moreover, stakeholders can help with project steering and subject matter expertise.

172. B. The work breakdown structure is a deliverable-oriented decomposition of a project.

173. D. In a functional organization, the authority resides with the functional manager, not the project manager.

174. B. Finalizing project work and closing all vendor contracts take place during the Closing phase. Performing quality assurance would take place during monitor and control, and managing stakeholder expectations should occur throughout the project.

175. A. The project sponsor declares the project is complete. The project manager is not the critical role in this answer.

176. A. When requirements are changing, an Agile approach allows the organization to be readily able to adapt to the environment.

177. D. Monitoring the risks and issue log is an activity of the Execution life cycle phase since the majority of the work, and therefore the issue and risk probability, occur during this phase.

178. A. Best practices and internal processes are known as preferred or soft logic from a dependency standpoint. The three types of dependencies are mandatory, discretionary, and external. Therefore, internal would not be the correct answer, as tempting as it might seem.

179. C, E. The issue log and action items are created during the Execution phase.

180. B. The plan lays out the communication methods as well as the frequency.

181. B. Projects can have business processes just like functional units. This was a change to how a project's business process works.

182. D. The project management plan consists of all project planning documents, including the charter, scope statement, schedule, communication plan, and more.

183. B. The scope statement is a reference for what the product needs to do, what assumptions and constraints have been identified, and what key performance indicators will be used to track and measure the project's success.

184. A. The project manager is forcing the actions of the team member to get compliance. This is not the best technique for team building but may be necessary in temporary situations.

185. A. Make sure each issue has an identifier, an owner, and a due date to resolve to help keep the project on track.

186. A. The Closing phase shuts down the project by gaining formal acceptance of project deliverables and then turning over needed activities and service support to ongoing maintenance and support teams.

187. B. Expenditure reporting gives feedback in terms of where and how money is being spent on the project. The project manager can then track this feedback against planned spending and against the rate of spending.

188. B. Avoiding never results in a resolution and is potentially the least effective technique as nothing gets resolved.

189. C. Although the two companies could merge, the probability that they will do so would yield a score between 0.0 and 1.0 in assessing its likelihood.

190. A. This is an example of negotiating. The use of a third party, like the project manager in the question, can help in producing an outcome, and their neutrality can assist in reaching agreement.

191. B. An item moves from the risk register to the issue log when the risk is triggered. A risk becomes an active issue when it is triggered. For instance, if the cost of material starts to rise, it might trigger a budget risk that gets moved to the issue log to be actively managed.

192. C. Level 1 of the WBS is the project level. The first level of decomposition can commonly be the second level of the WBS (deliverables, phases, projects).

193. B. They are known as quality gates. Predefined acceptance criteria help ensure quality and minimize work. Quality gates can be used in both the Agile and waterfall methods.

194. B. Project team members are responsible for producing the deliverables spelled out in the project charter and scope statement.

195. C. A constraint is anything that limits the options of the project team or dictates a specific course of action.

196. D. Subject matter experts have expert judgment and experience in the topic or project area that should be consulted.

197. A. Task completion is communicated to the project manager and/or coordinator so they can signal the next activities to begin.

198. A, C. Communicating changes is a key function of status meetings so that expectations are communicated, input can be gathered, and updated assignments can be given. Issues are discussed so those decisions can be made and actions discussed.

199. D. A project manager spends up to 90 percent of their time communicating, and coordination during the planning of a project is reliant on high communication.

200. B. The business case helps to provide justification as to why resources should be assigned to a project and what the return could be on the project.

201. A. To achieve drastic a so change in quality might require the project team to look at all aspects of the business: hiring, training, processes, and materials to achieve the CEO's goal.

202. C. Key performance indicators are the critical and quantifiable measures needed to track success factors for a project, product, or organization. There should be a limited number of "key" indicators that need to be tracked.

203. D. It is critical to create a record of meetings, especially action items and decisions that were made so there is a record for lessons learned and audit purposes.

204. C. The project charter will also contain high-level information about assumptions, constraints, and risks. Remember, it also the formal authorization to begin project work.

205. A. The team member is not given a chance to tell their side of the story, and the project manager does not inquire as to why the behavior is happening. These solutions can be short-lived.

Chapter 3: Tools and Documentation (Domain 3.0)

1. C. Named after the American engineer and management consultant, a Gantt chart allows the users to see at a glance information about various activities, activity start and end dates, duration, activity overlaps, and when the project will start and end.

2. D. Collaboration tools allow for instant messaging, sharing work screens with each other over the network, and other areas, including videoconferencing. A wiki page might be used as a collaboration tool, but the collaboration toolset is much broader than just wiki pages.

3. A, D, F. A RACI chart is a matrix-based chart used to identify roles and responsibilities on a project. It stands for responsible, accountable, consulted, and informed.

4. D. A change request log can be a simple spreadsheet that has the ID number, date, description, and so forth of the change.

5. A. A process diagram is a visual representation of the steps in a process. Other names for process diagrams are process maps and workflow diagrams.

6. C. The project management plan is the collection of plans created to address budget, time, scope, quality, communication, risk, and other project elements.

7. B. The issue log helps to manage items that need to be monitored and/or escalated to minimize the impact on the project team.

8. D. The project organizational chart helps to clarify involvement on the project and can be used to help create a decision-making matrix indicating who has authority to make certain decisions.

9. D. Standard tools like Microsoft Project or Primavera, plus an exploding market of cloud-based tools, help to automate the creation and updating of the project schedule. More simplistic projects can use Microsoft Excel or a comparable tool.

10. A. Process diagrams are often called workflow diagrams, which show all the steps from start to end in a process.

11. A. Especially valuable at the beginning of a project, a run chart helps identify information about a process before there is enough information to set dependable control limits.

12. C. A project dashboard is a form of a status report that displays summary information from lots of different areas of the project to present a snapshot of key information for analysis and discussion.

13. C. The resource conflict with the needed aircraft is an issue that needs to be actively managed, so this would get added to the issue log.

14. C. The organizational chart would have a breakdown of the hierarchy for the project and show where the assignments are for each project function and resource.

15. B. Some projects become impossible to manage without the aid of software to help keep track of information and automate the various reports the stakeholders require.

16. D. Collaboration tools would allow screen sharing, joint document editing, video calls, and joint document sharing, task lists, and calendars.

17. C. A process diagram can be used in two forms: mapping the processes as they are, and then reworking them to how they should be.

18. D. This document is showing values from a variety of different areas on the project to show a balanced view of the project.

19. A. A project dashboard is a form of a status report that will display conditions of different data elements and key performance indicators to communicate the progress and challenges of the project.

20. B. According to *A Guide to the Project Management Body of Knowledge* (*PMBOK® Guide*), Sixth Edition, a histogram is a special form of a bar chart used to describe central tendency, dispersion, and shape of a statistical distribution.

21. D. A scatter chart can also be called a scatter diagram and is used to help identify a correlation between the dependent and independent variables.

22. A. Use of software can increase the amount of time during setup but can save lots of time in managing the project in later phases.

23. B. A Pareto chart shows the values in a bar chart in descending order from left to right, and then adds a line chart to show the cumulative score. This will allow the analyst to see which issues are causing 80 percent of the disruption and focus on them first.

24. B. Most likely, this analyst is using a histogram, which would be used to show the frequencies of problem causes to help understand what is needed for preventive or corrective action.

25. C. Project scheduling software, especially for complex projects, can aid in the ease of tracking data and producing reports. Remember, there is a learning curve for all users whenever a new software tool is introduced on a project.

26. A. The organizational chart describes the project team member organization and identifies reporting structures.

27. B. A dashboard is either an electronic or paper report that shows lots of information in the format of a dashboard display with different "instrumentation panels" showing key performance information from different project areas.

28. B. A run chart, or a run-sequence plot, creates a graph that displays observed data over a period of time.

29. B. Famous sites like Wikipedia are examples of simple web tools that allow users to create content for websites. Often, content can mean the difference between a popular website and one that doesn't get much web traffic.

30. A. When schedules do not align, and there is a need to share a detailed message about what is going on, what is being done to address an issue, and what alternatives are available, an email would be the superior way to share the information. The next day, it would also be okay to follow up with an in-person meeting to make sure the boss received the information.

31. A. Enterprise social media would allow for broad notices to go out to the general public, including traditional forms of media to let everyone know of the work and closed highway. An enterprise social media strategy helps to build a more substantial brand presence for an organization.

32. A. The burndown chart is a visual representation and measurement tool showing the completed work against a time interval to forecast project completion.

33. A. A review by an internal team, such as security or even an internal audit division, is an audit with a recommendation to fix the finding.

34. C. A budget burndown chart is a tool used to see how much budget is remaining on the project and how much time there is to spend it. Often using graphical representation can help convey the status of a project better than a written report.

35. A. This is an example of a PERT chart. A PERT chart is a graphical depiction of the project's timeline and helps to identify task dependencies.

36. A. Process diagrams create visualizations of the different steps needed in any process, typically created as a flowchart.

37. C. This is a milestone chart, which marks the most important steps in the project and the dates they are targeted to be completed.

38. B. Whiteboarding apps try to simulate the power of using dry erase markers on an actual whiteboard helping to brainstorm and communicate ideas.

39. D. A fishbone diagram, also called an Ishikawa diagram, is a visualization tool for categorizing the potential causes of a problem in order to identify its root causes.

40. C. Mitigation is associated with the prediction and response to risk, whereas an issue log tracks active incidents. A correct answer could be tracking what action is to be taken.

41. B. A histogram chart can be used to show the frequencies of problem causes to help understand what is needed for preventive or corrective action.

42. A. Publishing a meeting agenda and disseminating it in advance allows for team members to be prepared, allows meetings to be on time and focused, and improves the use of team members' scarce time on a project. Depending on the type of meeting, agendas can be part of a printed packet of materials for the participants.

43. C. Text messaging allows Mickey's boss to get an update in the middle of another meeting, and if deemed important enough, allows them to step out of the meeting to get more information.

44. D. A whiteboarding tool can help in this process through drawing and other tools just like on a real whiteboard. Drawing a chart or a relationship can help stimulate ideas, visually leading to improved creativity sessions.

45. A, B, C. Technological factors, times zones, and interorganizational differences are examples of communication methods rather than triggers that would initiate communication to occur. Examples of communication triggers include audits, project planning, project change, risk register updates, milestones, schedule changes, task initiation/completion, stakeholder changes, gate reviews, business continuity response, incident response, and resource changes.

46. B. The issue log will keep a unique identifier for each issue along with a description of the issue, owner, and due date.

47. C. A Pareto chart contains both a line graph and bars where the individual values are represented in descending order by the bars, and the cumulative total is represented by the line.

48. D. Dashboards allow for the users to select which elements to display and be reported on. Note: For the exam, look for the electronic delivery of information, but remember nondynamic dashboards can be effective in meetings as one-page brief sheets.

49. D. A dashboard gives a snapshot of many different elements of the project in "instrumentation panels" on a screen. A dashboard does not have to be electronic, and a single-page, printed dashboard can be an effective communication method.

50. B. This is an example of an issue that should be captured in the issue log, and then the project manager should address the issue with the individual or the functional manager.

51. B. There can be one, and only one, person who is accountable per task in a RACI matrix.

52. C. Anyone working on or associated with the project can submit a change that is then reviewed by a change control board or other process. As such, an executive sponsor is not needed for a change request.

53. D. Also called Ishikawa diagrams based on the causality diagrams created by Kaoru Ishikawa, fishbone diagrams are a tool to show causes of specific events.

54. C. Especially valuable at the beginning of a project, a run chart helps identify information about a process before there is enough information to set dependable control limits.

55. A. The issue log is a list of issues that contains list numbers, descriptions, and the other attributes listed in the question.

56. C. Face-to-face meetings, or in-person meetings, are the superior choice for working on conflict resolution, giving performance appraisals, and working on team building.

57. B. The project scope that is cut back to operate within the new budget is the most likely impact to this project. The schedule would not be impacted as there is no call to lengthen or shorten the project time.

58. A, E, F. The basic communication model is sender-message-receiver. This model reflects how all communication exchange occurs.

59. C. Parametric estimating often uses a quantity of work multiplied by the rate formula for computing costs.

60. C. The addition of more resources, the modification of scope, or the adjustment of the timeline are examples of some of the changes that might be requested.

61. D. A regression plan, or rollback plan, will identify the steps and level of effort to return to the original state.

62. B. The three types are Most Likely, Optimistic, and Pessimistic. The formula is (Most Likely + Optimistic + Pessimistic) / 3. This calculation generates an average of these estimates, providing a single expected estimate to use for planning purposes.

63. B. The project management office offers standardization for a project but typically does not get involved in the details of individual projects.

64. C. The Pareto Principle implies that 20 percent of problems take up 80 percent of a team's time to deal with them. A Pareto chart is both a bar and a line chart that shows the largest concentration of values from greatest to smallest.

65. C. Using a similar, past project to develop a high-level estimate, this type of estimating can also be called an order-of-magnitude estimate.

66. C. It is known as actual cost. This may include both direct and indirect costs but must correspond to the budget for the activity.

67. B. The type is bottom-up estimating. By beginning at the work package level of the WBS, cost for each activity is calculated and assigned to that work package.

68. A. Top-down estimating, or analogous estimating, is where high-level project cost estimates are used by comparing to a comparable project from the past.

69. C. The value of the work completed to date compared to the budgeted amount. This number is expressed as a percentage.

70. D. The formula for calculating CV is: $CV = EV - AC$. Accordingly, the correct answer is $-2,500$ and the project is over-budget. A positive Cost Variance (CV) means the project is under-budget, and a negative CV means the project is over budget. $7,000 - 9,500 = -2,500$; therefore the project is over-budget.

71. C. CPI (cost performance index) is an important index, since it tells you the cost efficiency for the work completed.

72. D. The burn rate is how fast the project is spending its allotted budget, or how fast the rate money is being expended over a period of time.

73. C. If the team is not collocated, routine announcements and updates can be distributed via email as a weekly newsletter or something comparable. Some organizations use collaboration software like Microsoft SharePoint where there is a list of all updates posted as they are gathered and the audience is invited back to read the update.

74. A. Using an incremental naming convention is a great way to keep track of versions. Also don't forget to make updates within the document itself to capture the version and the changes made in that version.

75. C. Due to the compliance nature of this topic, the need to have formal documentation of this communication is needed. As such, both sending an email explaining the compliance requirement and asking for an email in return acknowledging receipt and understanding will help create an audit trail.

76. C. The Gantt chart is a type of bar chart that shows task duration and dependencies.

77. B. Program Evaluation and Review Technique (PERT) is a statistical tool used to analyze and represent tasks, not costs.

78. D. A key performance indicator to meet the publisher's target is to complete 1/12 of the books each day, which would be 8,333. Meeting this target would let Carl know they are on track to complete on time. Missing this number would mean Carl would have to change the speed of the process or add more personnel to add more shifts.

79. D. Audits are reviews to obtain evidence to ensure good practices are followed, that the project team is behaving in an ethical and legal fashion, and to ensure there is no fraud, waste, and abuse occurring.

80. D. The Scrum retrospective is a form of a lessons learned meeting used in the Agile methodology to help improve future sprints.

81. A. The question implies a synchronous discussion needs to take place and video would not be an option because of the technological challenges lists. Therefore, a voice conference call (telephone) would be the best solution.

82. B. Time-tracking tools allow team members to record their time against different work assignments. This allows the project team to compare the actual work effort needed against the planned amount of time they thought the task would take. Time-tracking tools provide visibility to the project manager to ensure the team is working on the right tasks in the right order.

83. D. A task board will give the team the picture of the work needing to be completed and its status. This can be a physical board if the team is collocated, or an online application.

84. D. A requirements traceability matrix maps and traces user requirements with test cases. It consolidates into a single document all requirements proposed by the client and helps validate all requirements are checked via tests cases. This tool helps verify all needed functionality during software testing.

85. C. Workflow and e-signature programs are a major innovation to physically routing documents for signature as it allows great speed and visibility to where in the approval process the document is currently in. There may be additional rules in your organization for formal signatures if there is no program to validate the identity of the signatory.

86. A. Word processing tools help in the creation and formatting of documents. While it is possible to create additional tables in a spreadsheet or a chart or diagram to be included in the document, the word processing tool is the superior program to bundle all of the information together.

87. B. A spreadsheet is a basic form of a database and allows the use of functions for formatting, counting, and summing for analytical review.

88. C. A ticketing system helps keep track of users' issues and helps monitor the follow-up and repair of the system as well as communicate status back to the customer.

89. D. A charting or diagramming program like Microsoft Visio helps with stencils and templates for the creation of workflow diagrams, organizational charts, and fishbone diagrams. There are many programs available for such work.

90. C, D. A virtual meeting is where everyone is not in the same location, and the use of technology allows the meeting to occur.

91. D. The content of the message will often cause you to tailor the method of communication.

92. B. Recognize the different time zones/schedules being used. If your project spans multiple time zones and/or there are a variety of schedules being used, take these factors into consideration in scheduling the meeting.

93. B. Version control tools help keep track of what state the document is in, who has reviewed it, and what changes were made or recommended, and keeps older revisions for archive purposes.

94. C. People retain more information when they both see and hear data points. Presentation software when used properly can amplify the effectiveness of the story and message.

95. B. Calendaring tools help everyone to share their availability and sign up for a meeting time that works for them. This can be useful especially if the project requires lots of individual interviews and saves time by giving the open timeslots and letting the receiver choose the best time for them.

96. D. File sharing programs like Google Drive, Dropbox, and Microsoft OneDrive can help to store project information in one location and allow multiple members to see the documents. These tools may also have version control built-in to allow multiple authors to check documents out to maintain document integrity.

97. A, C. Licensing costs can also be a contributing factor in determining which approach the team takes to on-premises versus cloud solutions. If there isn't a need to share information across geographic locations and only a small few making edits to the files, a local implementation may make sense. If the project is a part of a larger program or portfolio, then a cloud implementation may make the most sense even if the size of the project is relatively small.

98. A. Process diagrams are often called workflow diagrams, which show all the steps from start to end in a process.

99. B. The risk register is the tool used to document all potential risk for future analysis and evaluation.

100. D. A scatter chart pairs numerical data to help the analyst look for a correlation or relationship.

101. D. Run charts often represent some aspect of the output or performance of a manufacturing or business process.

102. D. Plotted points on a scatter chart indicate the type of relationship, or the absence of a relationship, between two variables, one dependent and one independent.

103. A. Wiki knowledge bases allow users to freely create and edit web page content using a web browser.

104. C. Collaboration tools like videoconferencing, screen sharing, and sites that allow for the joint sharing of documents, tasks, and calendars can be invaluable to a project.

105. A. The hope of the marketing campaign is that some of their videos will go viral where it is repeatedly shared and viewed based on virtual "word of mouth."

106. D. The 3.5 mile per week target is a key performance indicator to help track the success of a particular project outcome.

107. B. In the IT world, software audits are routine to see if there is any true up or true down that needs to happen after a large project is complete.

108. A. A task board is a visual representation of the status of different tasks that need to be completed during a sprint or project phase.

109. C. A requirements traceability matrix can include additional information such as who designed the test case, whether user acceptance testing has been completed, defects found and their assigned ID, and who has ownership of the test.

110. B. An application that helps with templates and stencils in the creation or a root cause diagram would be the best choice. Depending on the complexity of the chart you are creating, some presentation software packages will have a simpler diagramming capability.

111. D. The risk register can contain additional elements, including the response plan or where the plan may be located.

112. C. The organizational chart will spell out who reports to whom and what teams might exist within the project.

113. C. Project scheduling software can help keep track of start and end dates, keep track of durations, produce Gantt charts, and track progress through the project.

114. A. To help determine whether the project should move forward, a status report would give a snapshot of the work done to date, the status of budget and schedule, and issues facing the project.

115. B. Depending on the analysis supporting the risk response plan, sometimes action may not be taken at all. Other times, a deliberate action is called for to deal with the risk.

116. C. Risk triggers are a sign that a risk event is about to occur or has occurred and that action is needed according to the risk response plan.

117. C. Wiki pages can be edited though a web browser and do not require special software. It would allow the three managers to add to a single page to keep a log on events and progress.

118. C. The addition of more resources, the modification of scope, or the adjusting of the time-line are examples of some of the changes that might be requested.

119. B. Notification of passing an inspection would be a milestone. Do not confuse it with task completion because the task in this example would have been wiring and terminating the electrical in the building.

120. C. A milestone chart helps track the important achievements during the course of a project. The chart helps signify key events or tasks' completion date and status.

121. D. For an impromptu meeting where the team is dispersed, a virtual meeting like a conference call would be preferable so that the team in the field can remain on-site and get back to work when there is a resolution. Moreover, the problem in the field might need someone on-site for observation to report to the team.

122. B. When there is a preset project end date, like when a law sets when an activity must be completed by, schedule compression can be used to shorten the project schedule.

123. D. Instant messaging allows quick back and forth with minimal distraction to others in attendance at the board meeting.

124. A. To help keep a project on schedule, the key is to minimize disruptions to project work time. As such, a voice conference or other type of virtual meeting will allow the team to participate in the impromptu meeting without stopping work to travel, deal with traffic, and park.

125. B. A Gantt chart is a type of bar chart that illustrates a project schedule.

126. A. The Cost Performance Index is EV / AC. Therefore 900 / 1100 = .82.

127. C. Calendaring tools can help improve efficiency in finding availability where individuals can attend a meeting or event.

128. C. The risk register will become more detailed through progressive elaboration as the project matures.

129. B, D. The correct answer would be any form of instant communication such as radio communication, calling a mobile phone, or text messaging. The communication plan should direct the method of communication used when corresponding with members in the field.

130. A. Meeting minutes are used to provide a recap of what was discussed in a meeting.

131. D. What did I accomplish yesterday?, What will I do today?, and What obstacles are preventing progress? are the three questions.

132. D. Meetings agendas can be valuable if the project team is struggling to stay focused or struggling to get to needed subjects in the limited time available to meet. Agendas can help improve the efficiency of meetings.

133. C. Action items are comparable to a to-do list of activities that need to occur for product completion. Remember this is a dynamic list that can often grow during the project, specifically in meetings.

134. C. This is a method of the Agile methodology to create a list of all the activities that need to be completed, whether they are included in the next sprint or not.

135. A. Conferencing platforms like Slack, Microsoft Teams, and Google Meet allow for instant messaging, screen sharing, file sharing, and video and voice calls.

136. A. A ticketing system is a form of a case management system that allows the user and the technician to track and report on the progress of the fix of the user's issue.

137. D. Cloud tools like Slido are easy to set up, make a meeting more dynamic, and hold attendees' interest by having a purpose for this phone or computer. These tools allow for real-time polling of the attendees' opinions and can be a great tool to draw out the participation of introverts who feel more comfortable with typing an answer than speaking out loud.

138. C. Real-time, multi-authoring tools like Microsoft Office 365 and Google Apps will not only allow several users to make changes in the same document at the same time, but they will also show the work of the other individuals and keep track of who is making what change for review and discussion later.

139. B. Voice conferencing, or teleconference, would be the ideal choice as it would not require Greg to travel, allows for back-and-forth communication, and wouldn't be interrupted by poor internet quality.

140. A. A run chart is a graph that displays observed data in a time sequence.

141. B. The formula for calculating cost variance is Earned Value – Actual Cost. Plug in the numbers as follows: $2,500 – $2,275 = $225, signifying that the project is under budget.

142. C. Vendor knowledge bases can come in two forms: those accessible to all web users, and those where only company representatives can access the information. In either case, they help the user to self-serve to try to solve a problem without calling a help center.

143. C, E. Action items are a tactical item used to record a to-do list in the course of meetings, and a request for proposal is a procurement vehicle to help obtain goods and services. The procurement plan might be a part of the project management plan, but the RFP would ultimately be added to archived project documentation.

144. B. A purchase order is a legally binding agreement that guarantees the delivery of goods/services along with the agreed upon compensation.

145. B. He should use an RFI. When there is inadequate information to decide to pursue a purchase, a request for information (RFI) can help the buyer gather information to make an informed decision. A RFQ is not an appropriate response in this scenario because Hugh's company is not ready to purchase where a time-limited quote would be useful. They are only investigating at this point.

146. C. In order for a vendor proposal to be accurate and meaningful to the organization who publishes an RFP, there must be a complete written explanation of the goods or services that are needed. This is known as a statement of work (SOW).

147. C. An RFI will help give information to WigitCom for them to make a determination as to what solutions are available, give data on the vendor who can help, and give an order of magnitude estimate to the costs.

148. B. A RFQ is a tool used to help determine estimates for time and cost through bids for specific products or services.

149. B. An organization issues an RFP when they are ready to begin work and are ready to procure goods and services in support of the work.

150. C. Scheduling software can aid in the tracking and reporting of tasks, schedule, and critical path.

151. D. A decision tree is a technique for making decisions when the consequences of a decision are uncertain by helping to deduce the most beneficial option.

152. C. The project is under budget. A CPI over 1 means that the project is spending less than was forecasted for the measurement date.

153. D. Key performance indicators can be used at different levels of a project or organization to help determine the success of reaching particular outcomes.

154. A. There can be both internal and external audits; either one should be independent in authority and practice.

155. C. A status report provides high-level detail on the progress and challenges the project is experiencing. The cadence, or frequency, of the report will be set by the communication plan.

156. B. Electronic signatures speed up approval processes and give visibility of who still needs to sign off on the document. Depending on the organization's licensing, this legally binding alternative can be used for both internal and external approvals.

157. B. Use of dynamic interactive tools like real-time surveys can amplify the effectiveness of meeting by helping to hold attendees' interest and can be a great way to capture opinions and comments for use later in the project.

158. A. Real-time, multi-authoring editing software allows word processing, spreadsheet, and presentations to be made through group participation simultaneously rather than having to pass a file off to a new author to work on by themselves.

159. C. As Jeff has completed the SOW, the organization knows what they need to procure and are ready to move forward. Therefore, this scenario describes a request for proposal.

160. A. An RFI will give a sense of the number of providers or contractors who can provide the goods and services in question. An RFI will also help to give a ballpark estimate for the costs.

161. A. When an organization needs help in figuring out what is available in the marketplace, they release a request for information (RFI).

162. B. RFQ generally means the same thing as IFB (invitation for bid). An RFI is similar to an RFQ in that they both serve the same purpose, and most organizations use one or the other of these methods to help determine estimates.

163. B. RACI stands for responsible, accountable, consulted, and informed, and is used to help clarify roles and responsibilities. This is not a procurement vehicle.

164. B. A memorandum of understanding is used when a legal agreement can't be created between the two parties.

165. C. A nondisclosure agreement ensures that what's discussed, revealed, or created is kept within the organization.

166. C. This process includes submitting a SOW, receiving bids from vendors and suppliers, evaluating proposals, and making a selection.

167. B. A bidders conference typically occurs right after an RFP is published to help improve the quality of the proposals submitted. Moreover, vendors can get a better sense of whether or not they have the information they need to decide to provide a bid.

168. A. As outlined in the contract, the seller can recoup costs that are allowable in the contract terms. This also allows flexibility to the buyer when the final outcome is not well-defined.

169. A. Fixed-price contracts are good agreements when the statement of work is clear and concise.

170. D. An RFI and an RFQ are both similar in the gathering of data to help an organization make an informed decision. The differentiator is the agency's desire to move forward with a purchase once the information is gathered and analyzed.

171. C. As the seller of the service continues to work on a project, the more chances for certain costs outlined to be reimbursed are increased. This becomes a trade-off between flexibility and uncertainty.

172. B. A fixed-price contract is risky to Wigit Construction as the seller because if there are problems on the project and it takes longer to complete, Wigit Construction must still pay for the labor and increased costs of materials to fulfill the contractual obligation.

173. C. Version control allows for a trail of what changes were added by who at which point in the document's history.

174. A. Early in a project is when risks are most likely to occur and accordingly can have bigger impacts on the project.

175. D. A fishbone chart is used to help determine and define the different factors that could be the cause of a problem and do analysis to determine the root cause.

176. B. Key performances indicators are a performance measurement tool to help track progress toward a goal or objective.

177. A. A task board is a visual tool used to create a picture of the work to be completed and the state the work is in. A task board can be either physical or virtual.

178. D. Specific actions that need to get completed get added to the action items list. This list is great during meetings to keep an accurate track of the tasks that come up during discussion so they can be successfully undertaken and completed.

179. D. The SOW specifies in detail the goods and services the organization is purchasing from outside the organization. This is a common prerequisite to the issuing of an RFP.

180. D. A request for information can have a similar meaning to a request for quotation in some companies. A RFQ can also have a cost quote that would be valid for a certain period of time.

181. A, D. RFIs and RFQs may be used interchangeably but may also have different meanings in different organizations. Ask questions to learn what these terms mean in your organization.

182. A. A request for proposal (RFP) is issued when an organization is ready to begin work and needs to procure goods and services.

183. C. Service-level agreements spell out the expected performance levels to be followed and adhered to. For instance, an electrical utility may have a requirement with their customers to restore power within a certain period of time in case there is an outage.

184. D. A nondisclosure agreement will spell out the confidentiality expected on the project, given legal protections to the company.

185. D. Time and materials contracts set an hourly rate for a contract worker. *We are here for you!* would get compensated only for the hours the two admin assistants work for the project/organization.

186. A. A purchase order creates the binding arrangement between buyer and seller, guaranteeing service to the buyer and compensation to the seller.

187. A. These are elements of expenditure tracking. This activity also includes measuring the project spending to date.

188. A. As outlined in the contract, the seller can recoup costs that are allowable in the contract terms.

189. D. Alterations in the way business is conducted, information is routed, and approvals are given are all business process changes.

190. A. Because WigitCom is paying for time and expenses and not for a defined deliverable, they will be able to change their mind as often as they are willing to pay for it.

191. D. Fixed-price contracts are good agreements when the statement of work is clear and concise.

192. A. Time and materials contracts set an hourly rate for a contract worker, such as a painter to finish an interior, instead of a fixed price. The total cost is not known at the outset and will depend on the amount of time spent to produce the product or service.

193. C. A fixed-price contract is risky to the seller because if there are problems on the project and it takes longer to complete, the seller still must pay for the labor and increased costs of materials to fulfill the contractual obligation.

194. D. Time and materials is tied to a rate and a number of hours. If the buyer is willing to forgo the task the contract worker was meant to do in favor of other work, or if the buyer would like to have additional work done at their cost, this contract vehicle allows that type of flexibility.

195. B. This is the activity that compares costs and expenses to date against the cost baseline so stakeholders can see the variance between what was planned for and what actually occurred.

196. C. A contract is legally enforceable and typically falls into three categories: fixed-price contracts, cost-reimbursable contracts, and time and material contracts.

197. C. The Wigit Construction bridge project is limited by time, budget, and scope, which are the elements of the triple constraints.

198. A. A request for information (RFI) is used when there is not enough information or expert judgment to know what a good or service will cost, or to understand how many vendors there are who can meet this demand.

199. C. As outlined in the contract, the seller can recoup costs that are allowable in the contract terms.

200. B. A histogram chart can be used to show the frequencies of problem causes to help explain what is needed for preventive or corrective action.

201. D. This is a project burndown chart.

202. B. With a fixed-price contract, the risk is on the seller. When changes to a project increase their costs but not their revenue, they stand to lose money. Accordingly, they would likely allow very few changes to the scope.

203. A. The project sponsor is looking at a single variable on the project, budget. That variable will probably not tell the complete picture on the project, including that the increased cost resulted from increasing the scope and not cost overruns. A balanced scorecard helps give that bigger picture.

204. C. Time and materials are great to use for staff augmentation situations. The customer is only on the hook for whatever time they use, so it gives the customer flexibility.

205. D. In terms of the triple constraints, the results would be the project taking longer with the resources that have been assigned to the project.

Chapter 4: Basics of IT and Governance (Domain 4.0)

1. D. Acceptance is choosing to live with the impact a risk would have on the project. Essentially, this means choosing inaction as the intentional decision on how to address the risk.

2. C. The risk response plan outlines the course of action that will occur should the risk materialize and start impacting the project.

3. B, C. Organizations will likely have policies and practices that govern the use of removable media and may outright prevent the use of these devices. In this question, Kevin's duty to protect the removable hard drive should include encrypting the device and making sure to physically lock up the device when not in use.

4. B. Remember that a risk can be either positive or negative and represents an opportunity that did not exist earlier in the project.

5. B. The risk register is the tool used to document all potential risk for future analysis and evaluation. It includes a list of risks containing the risk number, description, probability, impact, risk score, risk owner, risk response plan location, and more.

6. D. Risk analysis is the process of figuring out what risks might happen and what the results would be if that risk did occur.

7. B. Risk impact details the consequence or result that will occur if the event actually happens. Typically, this is categorized with a rating of High, Medium, and Low.

8. D. Risk probability is the likelihood that a risk event will occur and is expressed as a number between 0.0 and 1.0.

9. B, D, F. The techniques of brainstorming, interviews, and facilitated workshops can all assist in creating an initial risk list. The estimating techniques are used in activity estimation for a tasks, and a fishbone diagram would help create cause and effect relationships.

10. A. To share a positive risk, the project team seeks to assign the risk to a third party who is best able to bring about the opportunity. For instance, a company already has a working factory in a country where the materials are going to be sold, reducing costs on the project.

11. A. Expert judgment is where the opinions of subject matter experts, stakeholders, and project team members contribute their expertise from similar projects to help make a recommendation.

12. A. The probability and impact matrix is a tool used to calculate an overall risk score that can be used to help organize, prioritize, and determine which risks may require a response.

13. A. Teams not attending status meetings would be an example of an active issue, not a potential future consequence that will impact the project.

14. D. When dealing with a positive risk, to exploit it is looking for opportunities to take advantage of positive impacts. For instance, the price of oil drops and the project is able to purchase all the required oil at a significantly reduced price than was planned.

15. A, D, E. Standard information on a risk register might include an identification number, description, probability and impact, score, and owner as they are all related to risk.

16. D. Avoiding the risk attempts to bypass the risk all together including by eliminating the cause for a risk event.

17. B. Enhancing a risk involves monitoring the probability or impact of the risk event to ensure benefits are realized.

18. D. Risk identification is brainstorming and recording all potential risks that might occur during a project.

19. D. Risk tolerance is the threshold of comfort one has accepting a risk and its consequences.

20. B. There is interplay between time and cost on the project, meaning that scope will need to be adjusted accordingly.

21. C. The probability and impact matrix will also include a calculated final risk score that can be used to prioritize the risk's severity.

22. B. A risk trigger is an event that warns a risk is imminent and a response plan should be implemented. The formation of a hurricane that could make landfall and impact cost of materials is a risk trigger.

23. D. As the project progresses through the Execution phase, the project team would be looking for risk triggers and then engaging a risk response plan should the need arise.

24. C. While the two companies could merge, the probability would give a score between 0.0 and 1.0 assessing its likelihood.

25. A. Contingency reserves are used to cover unforeseen costs or time that were not identified as part of the planning process. It is important to set expectations that these funds are not for extra functionality or enhancements, but to deal with circumstances like cost overruns, variances in estimates, and specific risk situations.

26. A, D. Passive acceptance is when the project team is not able to eliminate all the threats to a project and decides not to take any action about the risk. Active acceptance includes developing contingency reserves to deal with risk should they occur.

27. C. The management team is looking for both the threats and opportunities the project might experience so that the appropriate risk response can be enacted.

28. A. Quality defects are issues to be managed through the project rather than risks. The other answers represent the triple constraints of time, budget, and scope as it pertains to quality.

29. B. SWOT stands for strengths, weaknesses, opportunities, and threats and is an exercise to do an environmental scan at a moment in time. This environmental scan can reveal risks to a project.

30. D. A regression plan, or rollback plan, will identify the steps and level of effort to return to the original state.

31. C. Mitigation of a risk is reducing the chance the risk would happen or lessening the impact in the event the risk does get triggered.

32. B. A nondisclosure agreement assures what's discussed, revealed, or created is kept confidential for WigitCom.

33. B. During any change where the team starts with a working product and the change disrupts the availability, it makes sense to reverse the changes. In some Agile practicing software companies, they may only roll forward to the next version of the software rather than reverse any changes. In this example, the availability of phones due to the security component would make it important to roll back.

34. D. NDAs are common contract vehicles for jobs in both the government and private sector that would require a security clearance.

35. D. All three of these roles can contribute to the identification of risk on a project.

36. C. When dealing with a cutting-edge technology project, secrecy can play a key role in being the first to market and/or protecting intellectual property. Confidentiality constraints would prevent the team member from sharing with the rest of the industry until the project is complete.

37. C. Internal reorganizations can disrupt a project if resources are no longer available to work on the project.

38. C. PII stands for personally identifiable information and is not part of ESG.

39. A. Companies and projects must be aware of applicable regulations and standards impacting products and services. Complying with privacy laws at both a national and a local level will be critical for the success of the project.

40. C. Awareness of company vision, mission statements, and values is key to help influence decision-making and behaviors of the project team.

41. B. The U.S. Department of Homeland Security (DHS) defines personally identifiable information or PII as "any information that permits the identity of an individual to be directly or indirectly inferred, including any information that is linked or linkable to that individual, regardless of whether the individual is a U.S. citizen, lawful permanent resident, visitor to the U.S., or employee or contractor."

42. B. Regulations and approvals can vary between countries and the various economic unions that can govern commerce both local and globally. Being aware of any regulations and standards that must be followed will help with the timely execution of a project.

43. C. HIPAA protected health information, or PHI, is any personal health information that can potentially identify an individual, that was created, used, or disclosed in the course of providing healthcare services, whether it was a diagnosis or treatment.

44. B. As more companies and industries make a shift to IT-driven services, understanding what standards are spelled out anywhere the service is used becomes a key factor of success. Failure to comply can lead to negative consequences anywhere, from large fines to restrictions for offering those services in certain countries or regions.

45. B, D. Sensitive personally identifiable information (SPII) is stand-alone data that by itself can identify an individual. If options A, C, and E are linked data and if combined with SPII, then they will become SPII.

46. D. DewDrops would be working to make sure the impact to the company brand value is in their control and that the project will not negatively impact their image.

47. C. When you know what you stand for, decisions become easier. In this example, there is a mismatch between the candidate's values and the values of the company. If the candidate were not trustworthy in representing their skills and experience, they wouldn't be trusted with patients' confidential information.

48. A. PII stands for personally identifiable information and is not part of ESG.

49. C. A customer relationship management (CRM) system is one that is used by companies to interact with their customers.

50. B. Transfer is the strategy of moving the liability for the risk to a third party by purchasing insurance, performance bonds, and other tools.

51. A. Measuring the Schedule Performance Index (SPI) and the Cost Performance Index (CPI) are a part of executing the project and not directly an activity of risk planning.

52. D. During the Execution phase, the project observes whether a risk is triggered and takes the planned course of action should that occur.

53. A. The strategies used to respond to positive risks include exploit, share, enhance, and accept. Mitigate is a response strategy used for negative risks.

54. A. The strategies used to deal with negative risks are avoid, transfer, mitigate, and accept.

55. D. The company is trying to enhance by monitoring the probability the risk will happen so they can act quickly to achieve positive benefits.

56. A. Since the need for success on her current project is probably needed for job security, Katie would most likely try to avoid risky decisions because the margin of error is so low.

57. D. Because of the stability within her job due to great performance, Karen probably has a bigger cushion to gamble and take more risks.

58. C. This strategy watches for and emphasizes risk triggers to help enhance the probability or impact of certain risks.

59. D. Quantitative risk analysis would add more concretely measured items like the numerical score for each risk.

60. A. The probability ranking in this analysis would be high, medium, and low, and getting expert judgment on what the impact would be.

61. D. The risk management plan outlines how the project will define, monitor, and control risks throughout the project.

62. A, C. Avoidance and acceptance are both risk responses to deal with negative risks.

63. A. The COVID-19 pandemic revealed the impacts travel can have in the spread of a highly contagious illness. Understanding how project needs can impact the local and global environment will influence the planning and execution of a project. In this example, you see various ESG factors in conflict with one another as DewDrops values in interviewing in person are in conflict with global safety. It is not uncommon for project managers to have to navigate conflicting constraints.

64. B. Maintaining datacenters represents the highest levels of cost, making IaaS the correct cost-saving option. Infrastructure as a service allows the cloud consumer to provision servers, storage, networking, operating systems, and other computing resources on demand. With IaaS, companies have the ability to stand up servers, operating systems, networking hardware, databases, and applications that mirror (or even improve on) the current datacenter environment.

65. A. A data warehouse is a collection of data for the purpose of helping stakeholders make sense out of the data for the purpose of analysis and decision-making. Data in a data warehouse may include internal as well as external data resources.

66. D. A platform as a service (PaaS) provider hosts the hardware and software on its own infrastructure for users over the internet.

67. A, C, D. Using a service will limit the control the customer has over features, updates, and timing of service. It is also possible that the costs will increase as the use of the service grows.

68. B. Reduced control can be a disadvantage to the project team and ongoing support when feature changes and timing of upgrades are controlled by the hosting company and not the customer.

69. A. This is the definition of infrastructure as a service, which provides virtualized computing resources over the internet.

70. D. Content management systems help with the organization's tagging, retrieval, and deletion of documents when a retention schedule has been applied to the documents.

71. D. Granted, this can be considered a bit of a trick question. Some organizations are poor at keeping current on an application, leading to risks related to doing no upkeep on the software. That said, being forced to update software and test against an organization's business process whenever the hosting company wants and regardless of the customer's timing is a challenge rather than a driver.

72. A. ERP systems help organizations manage day-to-day business activities and can vary in size and function depending on the organization and industry using the software.

73. B. Customer relationship management is a process organizations use to administer their interactions with customers.

74. C. XaaS is a collective term that refers to the delivery of anything as a service, in this example mobility as a service. XaaS provides a platform where users can access IaaS, PaaS, and SaaS providers as well as their on-premises services.

75. A. The Presentation layer ensures that the interactions passing through it are in the appropriate form for the data to be saved and used in other tiers of the architecture.

76. B. Multifactor authentication (MFA) is a security technology that requires multiple methods of authentication from independent categories or credentials to gain access to systems and networks.

77. A, D. Highly sensitive projects across industries may require the candidate to obtain a security clearance to gain employment. While common when working for certain government agencies, this may also be a requirement for companies that support those agencies. A background screening may be required for employment regardless of whether a security clearance is involved.

78. A. Organizations may lock down buildings or areas within buildings to make sure that only those individuals granted access may physically be in that location. This practice can protect again loss of trade secrets and unauthorized access by bad actors attempting to harm sensitive parts of a business. In areas of medical research, areas may be locked down to protect against contamination of clean rooms where research is being conducted or patients are being cared for.

79. B, C, E. The three pillars of access are the individual must have clearance, access, and a need to know the information. If they do not possess all three, the information should not be shared with them. For example, a team member might have clearance and access to information but in their role, they do not possess a need to know the information. Therefore, the information should not be shared with them.

80. B. The project is protecting trade secrets that will help them gain a competitive advantage in the marketplace. They are ensuring that they protect their intellectual property.

81. A. A patent is a limited property right relating to a discovery or invention in exchange for public disclosure of the invention. In this instance, DewDrops Medical needs to release information about their discovery so they can start selling the treatment.

82. C. A copyright grants the creator of the original work exclusive rights to use and distribute their creation. Copyrights are used in writings, music, and art, to name a few areas.

83. B. A trademark is used to protect a brand by identifying any word, name, symbol, device, or any combination intended to be used in the identification of goods and services and to distinguish them from other parties' goods and services.

84. B. There are multiple good answers in this question, but the best answer is to protect national security information. The clues in the question to look for are working with a national government, working with a department of defense, and would support operations in a battlefield. This is all information related to national security.

85. D. Removable media such as hard drives, USB drives, and even mobile phones are great for convenience but represent a risk of loss of data to bad actors when removed from a controlled environment. Moreover, it can violate data retention policies and the ability to find information if data is stored on a remote media device that is not plugged into the network.

86. A, C. While convenience and security are constantly in conflict with one another, the key reasons for this type of policy are to control protected data and to limit the risk to harmful software such as virus and malware.

87. B. HIPPA stands for the Health Insurance Portability and Accountability Act, which was enacted on August 21, 1996, and that governs standards for electronic exchange, privacy, and security of health information. This act defined protected health information.

88. D. The classification of PII and non-PII can be vague. Nonpersonal data such as generalized data, aggregated statistics, or a partially or fully masked IP addresses is data that will not allow you to identify an individual.

89. C. Change control can vary slightly between software and infrastructure. The key is to know what changes are made to the environment and the routines that would be followed if there is a problem with the change. This minimizes the amount of downtime the service has.

90. A. Maintenance windows are pre-negotiated time blocks that let the project team know when it is okay to bring down an IT service that is already in production.

91. C. A rollback plan is established ahead of time so that everyone knows what immediate action is needed to restore service within the maintenance window. In the middle of a process and or a crisis, it can be difficult to think on the fly the steps that might be needed to restore functionality.

92. B. Notifications happen before, during (depending on the length of the outage/change), and after. Always overcommunicate when you have a production change.

93. C. Validation checks help to ensure that the change was successful and that all services are available to the end user or customer.

94. A. A risk assessment happens when a change to software is requested to gain an understanding of all possible problems and opportunities that might exist with the change.

95. B. When the end-to-end process enabled by software is large and/or the changes are frequent, automated testing allows for better quality control before the application is released to production.

96. C. A tiered architecture allows for development and testing to take place outside of a production environment.

97. B. Manual testing is where a person and not a computer script or application performs testing through actions like logging in the system, running a report, or entering a record. This is appropriate if the application is small or the change is minor.

98. A. Knowing what success looks like can make or break a project and any change begins with a good requirements definition.

99. C. A change control manager or approval board helps to ensure everything is ready and the change has followed the appropriate process before a go decision is given.

100. B. Along with nondisclosure agreements, personnel actions are confidential and are not shared beyond the employee and management. As such, it can appear to those not in the know that nothing is happening to correct the problem.

101. D. When linkable data elements such as zip code, date of birth, and more are associated with an SPII data element, the linkable data element becomes sensitive personally identifiable information. SPII stands for sensitive personally identifiable information.

102. D. Facility access is a form of physical security used to prevent unauthorized access to a location, building, or part of a building.

103. A, C. Sustainable practices include many areas such as human rights, labor practices, the environment, fair operating practices, consumer issues, and community involvement. Option A acknowledges the need to preserve the environment and habitat for wildlife. Option C commits Wigit Construction to fair labor practices. Options B and D would violate the company's vision and values to be more sustainable.

104. C. Environmentally sound, socially responsible, and economically viable are three factors where trade-offs must be made in a similar way to the tactical triple constraints.

105. C. Multifactor authentication (MFA) helps protect personal information, financial assets, or company systems from access by an unauthorized third party. It is possible that each of the other options in this question could be one factor of authentication used in an MFA login.

106. C. While viewed in a silo, a focus on lowest-cost bidder alone could be detrimental to a buyer's vision, mission statement, and values. However, it is important to realize there still may be a business driver to focus on keeping costs low to stay on budget. Low costs are not a unique sustainability factor.

107. B. Time, budget, and scope are all technical issues related to project management. Awareness of applicable regulations and standards would be an environmental, social, and governance risk factor, or not technical to the core project itself.

108. B. Organizations and their project teams must be alert to the interests of the broader stakeholder community. In this instance, the regulations and standards of the new country will absolutely inform how the business will conduct business in-country. Moreover, in an era of global technology companies, one country's laws may force the entire company to adhere to their expectations such as in the cases of personally identifiable information collection.

109. B. While the GDPR acronym may not be on the test, it was one of the first major privacy laws and it informs how data is treated today. Knowing about the GDPR and its components can help you prepare for the exam and prepare you for when you are a project manager in the real world.

110. A. Financial services worldwide need to make sure they are PCI-DSS compliant to protect payment card handling and sort out vulnerabilities.

111. C. This question provides an example of a country-specific compliance situation. Being aware of this and similar laws will help provide context to questions on the exam. Moreover, it will help you prepare for real-world project management situations you might face.

112. C. Disclosure of information, unauthorized access to systems and information, and legal breaches would all be risks that failing to properly design and deploy information systems would have.

113. B. The work breakdown structure is characteristic of a more deliberate, up-front requirements gathering where many of the requirements are identified early in the project.

114. B. Physical security guides all parameters regarding where and how your data is stored. In addition to organization and project requirements, remember that different states and countries have very specific laws governing data security.

115. C. Operating in a different state or country than the project or product team members are accustomed to can mean your data storage practices do not align with the needs or legal considerations of your customer.

116. B, C, F. Smoke/fire alarms, physical access control systems, and data backups are all related to physical security. Network uptime and availability is a component of network security. User authentication would be application-level security, whereas risk management policies are compliance directives.

117. A. Data backup and availability is a component of physical security. Monitoring user activities and mobile application security are other areas of application security.

118. D. Reputational and brand harm resulting from a privacy breach can be severe and invite litigation, lead to a loss of business, or result in significant financial damages to the project and organization.

119. A. Confidentiality protects information from disclosure without the consent of the individual or organization the information is about.

120. B. A PIA is a risk management tool and can help evaluate risks and propose solutions to privacy and data confidentiality needs.

121. C. A background screen provides a wide variety of facts that can help determine the integrity and trustworthiness of a candidate. Disqualifying aspects for a position or to obtain clearance can include lying on an application, a previous criminal record, or significant credit or financial problems.

122. C. A security clearance is one of the three pieces related to data access: clearance, access, and a need-to-know.

123. A. Physical security requirements may restrict movement in and out of certain rooms, wings, or buildings.

124. C. Multifactor authentication, specifically for project team members and not users, is an example of information security. The other three options are all related to privacy and compliance identification and remediation.

125. D. A maintenance window communicates the time when a change to an IT system would be the least disruptive to the business. Maintenance windows can be set globally via a schedule or determined on an ad hoc basis.

126. E. Designing your project approach sustainably can have benefits across the board and will create sustainability benefits that survive beyond the end of the project.

127. C. Earning the right to do business in-country and avoid punishment is the most correct answer of these options. By proactively complying with the law, the company's brand value will not be adversely impacted, affecting its profitability and growth potential. Additionally, while some activities may be legal in certain countries like the minimum age of employment, it may be in the best interests of the company to comply with a higher standard than the law requires. Clearly, following regulations earns points as the right thing to do.

128. D. If you build a product and allow citizens from all countries to use it, then the GDPR will apply to the way European customers' data is processed.

129. A. Sustainability concepts in project management have emerged as one of the most important trends in project management around the world. Projects are the mechanisms that allow organization-wide change.

130. D. The three ESG factors of environmental, social, and governance become the triple bottom-line elements to sustainability relating to people, profit, and the planet.

131. B. The risk register is the correct location for the recording of ESG factors impacting, or potentially impacting, a project. It may be important to add a category field to the risk register to identify the ESG variables.

132. A. Projects must consider their impact on the environment, and identifying this impact is important as early as in the Discovery and Concept Preparation phase.

133. B. Projects must consider the impact on people whether it impacts their jobs, their way of doing business, or their community.

134. C. Awareness and adherence of applicable government regulations, industry standards, and organizational policies and procedures are a critical success factor of modern-day projects.

135. D. Companies who embrace sustainability beginning at the project level will have a competitive advantage over those organizations that do not by reducing future negative liabilities, increasing brand value, and attracting and retaining top talent.

136. B. Impacts on employees, clients, and others oftentimes take a backseat to financial, governance, and environmental concerns.

137. B. These are all elements related to social responsibility in project management.

138. A, C. Properly recycling electronic equipment helps environmental sustainability while also following local regulations.

139. B, C. Not accounting for appropriate approval process through government agencies will negatively impact a project schedule. Additionally, if the final product does not comply with pertinent regulations, it will add cost and delay the project in order to align to the standards.

140. D. While time is of the essence in most projects, a sustainable organization would look to balance the timeframe with its ESG goals.

141. C. It is common for assumptions and constraints to shift during a project, and when there is a major change linked to a brand's value, the correct course is to evaluate the impact of the change and communicate to the project steering committee any impacts to time, budget, and scope.

142. C. The principle of least privilege provides guidance on how to keep a project's or an organization's data safe. In this example, the CEO should be granted access to only the relevant documents they need to perform their job. Blindly granting extensive access to data if the individual does not have a business justification will increase the risk of data loss or cybersecurity risk. Only grant the minimum access needed to be able to perform the job.

143. A. The business justification in this example is clear since there was already a role established giving access to data when Ellen was in the position. The project manager has a business need to see all documents, reports, and deliverables associated with the project or program they are working in.

144. C. Data classification helps to protect intellectual property, trade secrets, national security information, and privileged conversations between a doctor or lawyer and their client.

145. B. If there is a sound business justification and the necessary approvals match the security protocols, there is no problem granting the access. The next step needs to validate the request before permissions are assigned to the engineering data.

146. C. When an employee suddenly leaves a project, either voluntarily or involuntarily, it is vital that their access to project materials and systems be removed immediately. This protects the organization from data destruction, sabotage, or loss of trade secrets.

147. D. When implementing a change to a production system, it is important to minimize any downtime so the system is available. A rollback plan is a key step to quickly recover to an available status if there is a problem implementing the change.

148. D. New software or changes to applications must be tested prior to release. Manual testing is a key step that must be successfully completed before automated tests can be created and deployed.

149. B. RACI charts, maintenance plans, and communication plans will inform how and when to give the customer updates on system activity. A good rule is to overcommunicate outages and changes with the appropriate stakeholders.

150. C. When architecting a system that brings various corporate functionality together, an enterprise resource planning (ERP) system is often chosen.

151. C. A sustainable approach would be to ensure a sufficient income for project team members to allow team members to provide support for themselves and their families.

152. A. The General Data Protection Regulation would apply if data were being collected and stored on European citizens. While it would technically be possible to set up two different processes for European and non-European citizens, the most efficient software would follow the guidelines holistically, and the obligations are not overwhelming anyway.

153. A, C. A risk becomes an active issue should it materialize, and the appropriate risk response strategy should be implemented.

154. C. A tiered architecture would create stand-alone environments for production, testing, development, QA, and training protecting each of these activities from each other and maximizing uptime. A tiered architecture is more expensive and requires more staff to maintain, so there are trade-offs between cost and availability.

155. D. While it might be tempting to make the fixes directly in the production environment, having established environments for this work to be performed helps control not just this outage but future outages. Making the changes in development and then promoting the code through testing and production is the correct choice.

156. A. The ITIL framework is widely used in the IT industry and sets the model for IT service delivery, including incident, problem, request, and change management.

258 Appendix ▪ Answers to Review Questions

157. B. On-premises change control is based on ITIL, the Information Technology Infrastructure Library, which sets best practices in IT service management.

158. C. Cloud change control is guided by AWS Cloud Adoption Framework (AWS CAF) and others and builds on the foundations created in ITIL change management.

159. C. A cloud change control process would benefit greatly from a speedy and agile approval process. However, when an organization transitions from on-premises to a cloud approach, typically legacy practices are cumbersome and slow compared to cloud implementation needs.

160. C. Cloud environments offer greater agility, which facilitates speed as it pertains to change. Licenses and subscriptions are always available and setting up new environments, and rolling back changes can happen with minimal steps.

161. B. Traditional IT environments must account for the sunk costs (previous investments) in existing infrastructure and handle multiple steps to be able to scale out the environment.

162. A. Organizational change management using an approach like Prosci ADKAR focuses on people adjusting to and adopting new ways of doing business. It is important to know that ADKAR is a great method for individual change if one is seeking to adopt new habits.

163. D. Software release management helps to maintain the available and supportability of the existing production environment.

164. B. Release management enables successful releases to a production environment with minimal disruption and ensures future supportability.

165. D. The ability to shut down the road so that the paving work can begin is an example of a downtime or maintenance window where the production system, in this case a road, is not available for use.

166. B. The best answer would be to roll back to the last stable production instance of the database. Maintenance windows are typically negotiated in advance and are not at the discretion of the technician to modify them. Note: There are instances where the downtime can be extended with communication and approval.

167. C. Some organizations embracing rolling forward instead of rolling back utilize automation for build and testing for release to production.

168. A. Continuous integration asks developers to commit their code frequently to bring greater speed and agility to building new features and delivering the right products for their customers.

169. D. End-user or server-level computing refers to the hardware's capability to process programmed functions. There are several elements that go into determining hardware's computing power, including processor, storage, memory, network connectivity, and software version.

170. B. Storage is an IT term that can refer to both stored data and to the integrated hardware and software systems used to manage and secure data.

171. C. Databases are used to organize a collection of structured information and are typically stored in a computer system.

172. D. Networking and the connectivity it brings refers to the extensive process of connecting parts of a network to one another using routers, switches, and gateways.

173. C. Electronic document and records management systems allow for organization and storage of large volumes of documents and allow retention schedules to be created determining how long information should be stored and when to remove it from the system.

174. C. An ERP system such as Workday, SAP, or Oracle helps large organizations track and manage various day-to-day business functions. Summary data can often be offloaded to a data warehouse, but warehouses are not the primary systems for transactions like recording time and expenses.

175. A. Financial systems are configured with the organization's chart of accounts, which is a complete list of financial account codes used to help the organization categorize and track expenditures for planning and reporting purposes.

176. A. The key information in the question is storing the data where multiple stakeholders can analyze the data and drive future decision-making. A data warehouse stores cross-functional data for longer-term analytic purposes.

177. D. Software as a service is a strategy of most enterprise software companies. SaaS applications are typically accessed through a web browser and are typical for office software, messaging software, payroll processing, and more.

178. C. In complex organizations, documentation is key for organizational success, especially those whose primary focus is IT. Documentation helps communicate how something was created and/or configured and how to use the product or service.

179. D. Financial systems can be included as a part of an enterprise resource planning platform or exist as stand-alone applications. Financial systems serve as the single source of truth for all financial transactions and reporting.

180. B. CRM systems help support the customer engagement process to improve customer satisfaction by empowering the support staff with data and access to address customer concerns.

181. B. CMS tools are used to manage the creation and modification of digital content on a website without the need to write code from scratch.

182. C. Documentation is critical to understand decisions that were made regarding the creation of an application, how the software was configured, what changes were made over time, and how to use the product. Poor documentation leads to an inferior customer experience and service degradation over time.

183. A. The exam is looking to test knowledge related to electronic document and record management systems. It is important to understand that an EDRMS uses an organization's retention schedule to prevent premature deletion as well as the appropriate destruction of documentation. A document's retention may be set by law, regulation, industry standard, or organizational policy.

184. C. Anything as a service describes a general category of services related to cloud computing and remote access. In this example, it is a bucket for any end-to-end service that can be provided by a third party such as disaster recovery as a service (DRaaS).

185. A. Organizations' legal departments are becoming dependent on project management to achieve compliance with laws governing personal and protected information.

186. D. All these laws fundamentally ask businesses to change how they collect, store, and protect data while giving the consumer visibility into that data.

187. A. Project management serves at the crux of organizational change management initiatives. Accordingly, the project management discipline is core to keeping pace with compliance requirements as laws and regulations evolve over time.

188. C. As the use of the internet continues to evolve, just because a law is not affected in your country doesn't mean your project does not need to meet the compliance requirements. Moreover, it is common for regulations to get reproduced from country to country.

189. B. Noncompliant projects have major consequences both the consumer and the business. An increase in privacy protections is the key point of complying with laws, regulations, and IT security practices.

190. D. It is estimated that the cost of noncompliance is three times higher than the cost of compliance. Complying with the regulatory environment checks the box legally, but the business is also doing the right thing, the best argument to combat the sticker shock of complying is that the cost of noncompliance is higher.

191. B, C. Companies and projects that comply hold a competitive advantage over those that do not. Moreover, the number of lawsuits brought against companies for noncompliance is increasing liability and costs, and even risk a company's survival.

192. A. Many associations, such as the Project Management Institute, asks their membership to agree to follow a Code of Ethics and Professional Conduct. This helps instill confidence in the project management profession as a whole. More importantly, it helps aspiring project management individuals become better overall practitioners.

193. C. Ethical professionals make the best leaders, lead by example, and create an environment where project team members want to do the right thing.

194. B. The best practice is to include regulatory impact assessment at project initiation as it will set the mindset of compliance for the entire project. It can be difficult to make a significant mindset change when a project is in flight.

195. D. Compliance requirements and risks need to be ingrained in all projects and not bundled in a stand-alone and often symbolic project.

196. B. Regulatory impact assessments are a best practice and should be performed in good faith regardless of the repeatability of the project and whether it pertains to privacy. Some governments have regulations governing how tax dollars can be spent and event guidance on how projects must be completed.

197. B. This legislation enhances accountability and best practices in project and program management at the federal level. While a question on the PMIAA is unlikely on the exam, the question is a reminder that beyond open data and privacy, it is important to be aware of other industry compliance regulations that need to be followed during a career in project management. Performing an individual regulatory impact assessment could be a differentiator when applying for project management jobs and give you a competitive advantage.

198. D. Education is the first step in handling projects, creating systems, and securing data to make sure stakeholder information is correctly protected.

199. A. Individuals who are finished with the project but have not started a new assignment are benched resources.

200. A, D, F. When there is no harmony with the team, team-building exercises can assist in restoring morale. When a phase is completed, it can be beneficial to celebrate the progress on the project and recognize the contributions of the team. Any kind of change in the team dynamic can impact the project, especially when the leadership is new.

201. C. SWOT involves analyzing the project from strengths/weaknesses that are internal to the organization, as well as opportunities/threats that are external to the organization.

202. A. Cost of quality, or COQ, represents the costs needed to bring a substandard product or service produced by the project within the standard described by the Quality Plan or other criteria.

203. B, D. The risk score is the risk probability, typically expressed as a number between 0.0 and 1.0, multiplied by the risk impact. Projects may choose different quantifiable measures for probability and impact, and that would be spelled out in the risk plan.

204. A. This technique modifies the project schedule when accounting for unforeseen issues or resource availability.

205. A. The entire R&D industry would likely have a very high risk tolerance because they need to constantly challenge the status quo to become relevant.

Chapter 5: Practice Test 1

1. B, C. A project is an organized effort to fulfill a purpose and has a specific end date. Additional project characteristics are that it is temporary in nature, has a specific start and end date, and exists to provide a unique product or service.

2. A. Once accepted by the seller, this becomes a legally binding document for goods/services to be provided, and for compensation to be given at the listed price.

3. C. A scope statement documents the project's objectives, deliverables, and requirements, which are used as a basis for future project decisions. This is a more detailed explanation of what the project aims to achieve when compared to the project charter, which contains a high-level scope definition.

4.   A. NDAs are common contract vehicles for jobs in both the government and private sector that would require a security clearance. They are used to ensure that sensitive or trade secret information is not shared outside the organization.

5.   A, E. Sensitive personally identifiable information (SPII) is stand-alone data that by itself can identify an individual. If options B, C, and D are linked data and if combined with SPII, then they will become SPII.

6.   B. This is especially true when tasks are along the critical path, or when there is no float for that activity. Float is the amount of additional time a task has to complete before it begins to disrupt the critical path of a project.

7.   B. This process takes the more subjective results from the qualitative risk analysis and ties them to more numerical measures for evaluation and prioritization purposes.

8.   B. A Gantt chart is a type of bar chart that illustrates a project schedule.

9.   C. The "champion" role of the sponsor is very important both initially and as the project commences to keep the energy and focus of the whole organization committed to the project's success.

10.   D. The Cost Performance Index is Earned Value / Actual Cost. Therefore 800 / 1200 = .667. The project is over budget. A CPI under 1 means that the project is spending less than was forecasted for the measurement date.

11.   C. A Scrum meeting is typically held in the morning at the same time, in the same place, and it sets the context for the team's work.

12.   A. The statement of work (SOW) specifies in detail the goods or services that you are purchasing from outside the organization.

13.   B. ERP systems help organizations manage day-to-day business activities and can vary in size and function depending on the organization and industry using the software.

14.   D. This is a tool used to perform a moment-in-time environmental scan, looking both within and external to the organization.

15.   A. The Closure meeting will allow for the status of all activities, for the project team to return any property like ID badges/key cards and bring that phase of the work effort to an end.

16.   C. Software as a service is a strategy of most enterprise software companies. SaaS applications are usually accessed through a web browser and are typical for office software, messaging software, payroll processing, and more.

17.   B. Multifactor authentication (MFA) is a security technology that requires multiple methods of authentication from independent categories or credentials to gain access to systems and networks.

18.   B. Planned value is the total cost of work planned as of the reporting date, and it is calculated by multiplying the hourly rate by the total planned or scheduled hours.

19. D. Notifications happen before, during (depending on the length of the outage/change), and after. Always overcommunicate when you have a production change.

20. A. HIPPA stands for Health Insurance Portability and Accountability Act, which was enacted on August 21, 1996 and governs standards for electronic exchange, privacy, and security of health information. This act defined protected health information.

21. C. Organizations and their project teams must be alert to the interests of the broader stakeholder community. In this instance, the regulations and standards of the new country will absolutely inform how the business will conduct business in-country. Moreover, in an era of global technology companies, one country's laws may force the entire company to adhere to their expectations such as in personally identifiable information collection.

22. A. The key is actually setting money aside to deal with the risk should it occur during the course of the project. Failing to plan in this way can lead to project disruption or cancellation.

23. C. The addition of more resources, the modification of scope, or the adjustment of the timeline are examples of some of the changes that might be requested.

24. D. Milestones are used in project manager to mark a specific point along a project timeline, such as the project start and end date, completion of a phase, or gate checks.

25. A. A Pareto chart contains both a line graph and bars where the individual values are represented in descending order by the bars, and the cumulative total is represented by the line.

26. A. A Scrum master typically coordinates this meeting, which is a collaborative effort to detail all the work that needs to be completed during a work period (such as a sprint) and each item's respective acceptance criteria.

27. C. An RFI will give a sense of the number of providers or contractors who can provide the goods and services in question. An RFI will also help provide a ballpark estimate for the costs.

28. B. Financial services worldwide need to make sure they are PCI-DSS compliant to protect payment card handling and sort out vulnerabilities.

29. A. Audits are reviews to obtain evidence to ensure that good practices are followed, that the project team is behaving in an ethical and legal fashion, and that there is no fraud, waste, and abuse occurring.

30. C. When you know what you stand for, decisions become easier. In this example, there is a mismatch between the candidate's values and the values of the company. If the candidate wasn't trustworthy in representing their skills and experience, they shouldn't be trusted with patients' confidential information.

31. D. Sign-off on the project charter would be a milestone, not a deliverable.

32. D. Standard tools like Microsoft Project or Primavera, plus an exploding market of cloud-based tools, help to automate the creation and updating of the project schedule. Simpler projects can use Microsoft Excel or a comparable tool.

33. B. A histogram chart can be used to show the frequencies of problem causes to help understand what is needed for preventive or corrective action.

34. A. A change request log can be a simple spreadsheet that has the ID number, date, name, and description of the change.

35. D. The issue log constraints list numbers, descriptions, and owners of the issue, and starts to develop as the execution of the project gets underway.

36. B. During the kickoff meeting, formal approval for the project might occur in the form of project sign-off.

37. A. Much like the organizational breakdown structure, the RBS lists the different resources by categories such as labor, heavy equipment, materials, and supplies.

38. B. A statement of work formally captures and defines work activities, timeframes, milestones, and deliverables a vendor must meet in the performance of work for a customer.

39. D. On the exam, it is likely that there will be a scenario-based question that will test your critical thinking skills to reason through the correct activity sequencing of a project. This question should help give you a taste of what you might encounter.

40. A. With a fixed-price contract, the risk is on the seller. When changes to a project increase their costs but not their revenue, they stand to lose money. Accordingly, they would likely allow very few changes to the scope.

41. C. The communication plan documents the types of information needed by the stakeholders, when the information should be shared, and the method of delivery.

42. B. Know the difference between a merger and an acquisition. The latter is where one organization takes over the other one.

43. A. RACI stands for responsible, accountable, consulted, and informed, and it is a tool to help designate roles and responsibilities on a project. Conducting a RACI exercise can be one of the most valuable ways to contribute to reduced anxiety and clear expectations.

44. D, E, F. Time, budget, and scope are constraints in terms of how they affect quality.

45. A. By requiring the successful vendor to carry the insurance, it allows the consequences of negative risk to be transferred to the third party.

46. D. In the storming stage, the process of establishing who is the most influential occurs, and there is jostling for position.

47. C. Risk identification, probability and impact analysis, and risk response would all take place during the Planning phase of the project.

48. D. The business case can include justification, alternative solutions, and alignment to the strategic plan.

49. D. While stakeholders start out with a lot of influence, it decreases as the project advances.

50. A. A project is a success when stakeholder expectations have been met. This is the most critical factor involved when determining whether or not a project is a success.

51. B. A Gantt chart allows the users to see at a glance information about various activities, activity start and end dates, duration, activity overlaps, and when the project will start and end.

52. B. The project management plan is the collection of plans created to address budget, time, scope, quality, communication, risk, and other project elements.

53. A. A nondisclosure agreement ensures that what's discussed, revealed, or created is kept confidential by the various parties working with intellectual property, trade secrets, or classified information.

54. C. A data warehouse is a collection of data for the purpose of helping stakeholders make sense out of the data for the purpose of analysis and decision-making. Data in a data warehouse may include internal as well as external data resources.

55. D. A PaaS provider hosts the hardware and software on its own infrastructure for users over the internet.

56. A. In the data tier, information is stored and retrieved, and then passed back to the logic tier for processing. Eventually, the data is returned to the user.

57. A. As more companies and industries make a shift to IT-driven services, understanding what standards are spelled out anywhere the service is used becomes a key factor of success. Failure to comply can lead to negative consequences, such as large fines or restrictions for offering those services in certain countries or regions.

58. D. When implementing a change to a production system, it is important to minimize any downtime so that the system is available. A rollback plan is a key step to quickly recover to an available status if there is a problem implementing the change.

59. C. Project architects design plans and specifications for a project, such as in construction. Architects meet with clients and work closely to ensure understanding of requirements such as deadlines, budget, and structural needs.

60. A. A project dashboard is a form of a status report that displays summary information from lots of different areas of the project to present a snapshot of key information for analysis and discussion.

61. B. Using software can increase the amount of time during setup but can save lots of time in managing the project in later phases.

62. C. Especially valuable at the beginning of a project, a run chart helps identify information about a process before there is enough information to set dependable control limits.

63. D. Collaboration tools allow for instant messaging, sharing work screens with each other over the network, and other areas, including videoconferencing. A wiki page might be used as a collaboration tool, but the collaboration toolset is much broader than just wiki pages.

64. B. Most likely, this analyst is using a histogram, which would be used to show the frequencies of problem causes to help understand what is needed for preventive or corrective action.

65. A. The burndown chart is a visual representation and measurement tool showing the completed work against a time interval to forecast project completion.

66. B. According to the *Project Management Body of Knowledge (Sixth Edition)*, a histogram is a special form of a bar chart used to describe central tendency, dispersion, and shape of a statistical distribution.

67. B. A spreadsheet is a basic form of a database and allows the use of functions for formatting, counting, and summing for analytical review.

68. B, D. Active acceptance includes developing contingency reserves to deal with risk should they occur. Passive acceptance is when the project team is not able to eliminate all the threats to a project.

69. D. Monitoring the risks and issue log occurs during the Execution phase and are not an activity of the Closing phase.

70. A. Change control can vary slightly between software and infrastructure. The key is to know what changes are made to the environment and the routines that would be followed if there is a problem with the change. This minimizes the amount of downtime the service has.

71. B. Knowing what success looks like can make or break a project, and any change begins with good requirements definition.

72. D. This document is showing values from a variety of areas on the project to show a balanced view of the project.

73. C. It is a successor task because storyboarding would come after the development of the script. In the creation of an animated movie, many more steps than these are required to create the final product.

74. A. SIPOC-R is a process improvement method, and the acronym stands for Suppliers, Inputs, Process, Outputs, Customers, and Requirements. It is not used for cost estimating.

75. C. Creation of the work breakdown structure involves organizing the team's work by breaking it down into manageable chunks, or sections. This effort needs to happen before work begins and therefore occurs in the Planning phase.

76. B. The burn rate is how fast the project is spending through its allotted budget or the rate money is being expended over a period of time.

77. B. CapEx is important to help companies grow and maintain their business by investing in new property, equipment, products, or technology.

78. C. Examples of OpEx would include rent, office equipment, supplies, legal fees, travel expenses, and utilities, to name a few.

79. B. The prequalified vendor list helps buyers prescreen interested parties that wish to sell products and services. This can help create a pool of prospective sellers, which can speed up future procurement processes, establish relationships, and increase the chances a seller will be noticed and selected for future business.

80. B. That WBS components should always happen concurrently with determining major deliverables is not correct. Through progressive elaboration, or through many iterations, is where work packages are created. Major deliverables are created with the preliminary scope statement and included in the charter. It is highly likely the work needed to create those deliverables would be identifiable that early in the project before scheduling, team creation, and costing is completed. Identification of lower-level WBS components occurs after the major deliverables have been identified.

81. C. The lowest level recorded in the WBS is the work package.

82. A. This is an example of a PERT chart. A PERT chart is a graphical depiction of the project's timeline and helps to identify task dependencies.

83. C. The transition or release plan should be created in the Planning phase. Starting this work early allows for the end state of the project to be imagined for the transition to operations.

84. A, B, D. The Sprint goal sets the single objective and commitment of the developers. The backlog brings transparency to all work that is needed to be performed. The action plan for the sprint is dynamic and changes as the Sprint progresses.

85. E. It is vital to verify deliverables to ensure they meet specifications and satisfy project objectives. Additionally, vendor payments are often tied to deliverables, so before the vendor is paid and let off the hook for performance, make sure the deliverable is what the project expected it to be. Option B is scope creep that would prevent the closure of the project.

86. B. An activity of the Closing phase, the project manager validates all project expenditures, closes purchase orders and contracts, and finalizes the total spend for the project.

87. B. The project charter is prepared and agreed to in the Initiation phase of a project.

88. C. Established communication channels are likely to be created to ensure information governance, confidentiality, and information retrieval.

89. B. To help control costs and to minimize disruption to project work, a virtual meeting would be the ideal choice to handle routine meetings.

90. A. The formula for cost variance (CV) is CV = EV – AC. The calculations would produce a value of 2,500, indicating the project is under budget. A positive Cost Variance (CV) means the project is under budget, and a negative CV means the project is over budget. 9,500 – 7,000 = 2,500; therefore, the project is under budget.

91. C. As set in the project plan or communication plan, the retention of emails, instant messages, text messages, and other forms of communication should be established. It is likely the company has a retention schedule already established that would just need to be followed.

92. C. A requirements traceability matrix maps and traces user requirements with test cases. It consolidates into a single document all requirements proposed by the client and helps validate all requirements are checked via tests cases. This tool helps verify all needed functionality during software testing.

93. A. Daily stand-up meetings are typically used with an Agile methodology. As such, a daily stand-up meeting would not have a governance body in attendance and would normally be attended by project team members.

94. B. Hard timeboxing can be helpful in having a meeting move on against the desire for perfection or creative churning over possibilities.

95. D. Actual Cost may include both direct and indirect costs but must correspond to the budget for the activity. This can include labor costs, cost of materials, and use of equipment.

Chapter 6: Practice Test 2

1. C. Testers and quality assurance specialists inspect and monitor project activities and test products to ensure they are meeting quality thresholds and standards. These positions oversee the inspection and testing of products.

2. B. A kickoff meeting is a form of communication rather than a trigger of communication.

3. A. A risk can be either positive or negative and represents an opportunity that did not exist earlier in the project.

4. C. The industry plus the fact that the company is relatively new would suggest a more aggressive, gambling type of risk tolerance for the company to grow and succeed.

5. C. Constraints are those conditions that restrict the project in certain ways, like resources, timeframes, schedules, or budget.

6. D. When dealing with a positive risk, to exploit it is looking for opportunities to take advantage of positive impacts. For instance, the price of oil drops and the project is able to purchase all the required oil at a significantly reduced price than was planned.

7. B. Companies and projects must be aware of applicable regulations and standards impacting products and services. Complying with privacy laws at both a national and a local level will be critical for the success of the project.

8. C. Multifactor authentication (MFA) helps protect personal information, financial assets, or company systems from access by an unauthorized third party. It is possible that each of the other options in this question could be one factor of authentication used in an MFA login.

9. A. When architecting a system that brings various corporate functionality together, an enterprise resource planning (ERP) system is often chosen.

10. C. Short stories help focus on how the product is going to be used, which helps shape how the final product is designed.

11. D. There can be one, and only one, person who is accountable per task in a RACI matrix.

12. C. The project management office offers standardization for a project but does not get involved in the details of individual projects.

13. B, C. Highly sensitive projects across industries may require the candidate to obtain a security clearance to gain employment. While common when working for certain government agencies, this may also be a requirement for companies that support those agencies. A background screening may be required for employment regardless of whether a security clearance is involved.

14. D. Facility access is a form of physical security used to prevent unauthorized access to a location, building, or part of a building.

15. D. A change request log can be a simple spreadsheet that has the ID number, date, description, and so forth of the change.

16. B. Using an incremental naming convention is a great way to keep track of versions. Do not forget to also make updates within the document itself to capture the version and the changes made in that version.

17. D. A fishbone diagram, also called an Ishikawa diagram, is a visualization tool for categorizing the potential causes of a problem in order to identify its root causes.

18. C. The project scope being cut back to operate within the new budget is the most likely impact on this project. The schedule would not be affected, because there is no reason to lengthen or shorten the project time.

19. C. There are not enough resources for the tasks assigned, which leads to overallocation of the staff working on the project.

20. B, C, D. The ways to organize the WBS are by subprojects (where the project managers of the subprojects each create a WBS), by project phases, or by major deliverables.

21. B. The issue log will keep a unique identifier for each issue along with a description of the issue, owner, and due date.

22. B. Two of the roles of the project sponsor illustrated are the approval authority for the project and to help remove roadblocks in the project team's way.

23. D. As outlined in the contract, the seller can recoup costs that are allowable in the contract terms.

24. C. The project management plan is the collection of plans created to address budget, time, scope, quality, communication, risk, and other project elements.

25. B. In a balanced matrix organization, the project manager and the functional manager both control the budget and share power and authority.

26. A. A run chart is graph that displays observed data in a time sequence.

27. C. The Pareto principle implies that 20 percent of problems take up 80 percent of a team's time to deal with them. A Pareto chart is both a bar and line chart that shows the largest concentration of values from greatest to smallest.

28. C. Milestones can also represent the completion of major deliverables on a project.

29. B. The 95 percent phenomenon is where the project seems to be stagnant or unable to complete the last step or steps so the project can be completed.

30. D. Activities that monitor the progress of the project and require corrective actions occurs in the Execution phase.

31. D. Probability and impact is a prioritization tool and not a category that might be included in a risk breakdown structure.

32. C. A purchase order is an official offer issued to a seller indicating types, quantities, and agreed prices for products and services.

33. A. The team member is not given a chance to tell their side of the story, and the project manager does not inquire to why the behavior is happening. These solutions can be short-lived.

34. A. A checkpoint between project phases where approval is obtained to move forward. Usually the project reports to a steering committee to help ensure accountability on the project for time, money, and scope.

35. D. Scope creep is where the definition is constantly expanding and there starts to be little control on being able to finish the project.

36. B. The burndown chart is a visual representation and measurement tool showing the completed work against a time interval to forecast project completion.

37. C. Meeting minutes capture the details for why decisions were made, what was discussed, and any assignments/due dates that materialized out of the meeting.

38. A. Mitigation of a risk is reducing the chance the risk would happen or lessening the impact in the event the risk does get triggered.

39. A. A Scrum retrospective is a form of lessons learned done with the Agile methodology at the end of each sprint.

40. C. The Planning life cycle phase sets the details for how the project will be caried out and creates the majority of the project plan artifacts, including the project schedule, risk plan, communications plan, and other pertinent documents.

41. E. Other closing activities would include getting project sign-off, validating all deliverables, close contract, remove project team access, and conducting training as a part of the transition/integration plan.

42. C. The Planning phase includes identifying both what resources are needed and when they will be needed.

43. D. Forming, storming, norming, performing, and adjourning is the mode of group development that was developed by Dr. Bruce Tuckman.

44. B. Maintenance windows are pre-negotiated time blocks that let the project team know when it is okay to bring down an IT service that is already in production.

45. A, E, F. Commonly referred to as the triple constraints, almost all projects are constrained by time, budget, and scope as they impact quality.

46. D. Reputational and brand harm resulting from a privacy breach can be severe and invite litigation, lead to a loss of business, or result in significant financial damages to the project and/organization.

47. D. Physical security guides all parameters regarding where and how your data is stored. In addition to organization and project requirements, remember different states and countries have very specific laws governing data security.

48. A. Confidentiality protects information from disclosure without the consent of the individual or organization the information is about.

49. B. A PIA is a risk management tool that can help evaluate risks and propose solutions to privacy and data confidentiality needs.

50. A. The work breakdown structure breaks the project down into smaller, more manageable chunks of work.

51. D. Requirements are the characteristics of deliverables often represented in must have, needed, and wanted categories.

52. D. Capital expenses are used on spending for a benefit that will last longer than a year, and some companies might have a dollar amount that a single item might cost that would also lead to the purchase being qualified as a capital expense. As an example, a company might have a $5,000 limit on the individual cost of an item, meaning the purchase of a $6,000 server would qualify as a CapEx but a $2,000 desktop computer would not.

53. C. Terms of reference in a contract explain the objectives, scope of work, activities, tasks to be performed, and other information such as the structure of a project.

54. B. A return on investment (ROI) analysis can be used to determine if there will be a positive return on a project's investment.

55. A. A dashboard is a form of a status report that will display conditions of different data elements and key performance indicators to communicate the progress and challenges of the project.

56. C. Collaboration tools would allow screen sharing, joint document editing, video calls, and joint document sharing, task lists, and calendars.

57. B. During any change where the team starts with a working product and the change disrupts the availability, it makes sense to reverse the changes. In some Agile practicing software companies, they may only roll forward to the next version of the software rather than reverse any changes. In this example, the availability of phones due to the security component would make it important to roll back.

58. D. Wiki knowledge bases allow users to freely create and edit web page content using a web browser.

59. C. Voice conferencing, or teleconference, would be the ideal choice because it would not require team members to travel, allows for back-and-forth communication, and would not be interrupted by poor internet quality.

60. D. A scatter chart graph pairs numerical data to help the analyst look for a correlation or relationship. A scatter chart can also be called a scatter diagram and is used to help identify a correlation between the dependent and independent variables.

61. C. Timeboxing can be either hard or soft. Hard timeboxing dictates the task or activity must stop when your time is up, regardless of the status. Soft timeboxing is more flexible but gives the expected time a task should stop. This can be useful in complex tasks or activities.

62. B. Follow-ups check on the status of a new developments in the project environment that may have an impact on the scope, cost, or success of the project. These developments can be either internal or external.

63. A. DewDrops would be working to make sure the impact to the company brand value is in their control and that their project will not negatively impact their image.

64. C. Environmentally sound, socially responsible, and economically viable are three factors where trade-offs must be made in a similar way to the tactical triple constraints.

65. D. New software or changes to applications must be tested prior to release. Manual testing is a key step that must be successfully completed before automated tests can be created and deployed.

66. B. On-premises change control is likely based on ITIL, the Information Technology Infrastructure Library, which sets best practices in IT service management.

67. A. Datacenters consist of infrastructure, making IaaS the correct option. Infrastructure as a service allows the cloud consumer to provision servers, storage, networking, operating systems, and other computing resources on demand. With IaaS, companies have the ability to stand up servers, operating systems, networking hardware, databases, and applications that mirror (or even improve on) the current datacenter environment.

68. B. A request for information (RFI) is used when there is not enough information or expert judgment to know what a good or service will cost, or to understand how many vendors there are who can meet this demand.

69. C. Fast tracking is a schedule acceleration technique where two tasks that are scheduled in parallel are started at the same time.

70. B. Risk tolerance is the threshold of comfort one has accepting a risk and its consequences.

71. B. Dealing with a rapidly changing environment requires the ability to react to new information and feedback. The other options would not take advantage of the flexible, easily changing environment of the project.

72. A. A change control board helps vet and manage changes to the scope. A change control board, in conjunction with the change control process, will approve or reject changes to the scope of the project.

73. D. The project schedule determines start and finish dates for project activities on the project. It also sets the sequence and dependence of activities.

74. B. A risk becomes an active issue when it is triggered. For instance, if the cost of material starts to rise, it might trigger a budget risk that gets moved to the issue log to be actively managed.

75. D. When someone has a role of informed, it means they are informed after a decision has been made or a result has been achieved.

76. A. The project objectives component will explain the attributes of the product, service, or result of the project.

77. D. Run charts often represent some aspect of the output or performance of a manufacturing or business process.

78. C. Reduced control can be a disadvantage to the project team and ongoing support when feature changes and timing of upgrades are controlled by the hosting company and not the customer.

79. C. The project organizational chart helps to clarify involvement on the project and can be used to help create a decision-making matrix indicating who has authority to make certain decisions.

80. A. Process diagrams create visualizations of the different steps needed in any process, typically created as a flowchart.

81. C. A decision tree is a technique for making decisions when the consequences of a decision are uncertain by helping to deduce the most beneficial option.

82. C. The sprint planning meeting sets what can realistically be accomplished during the sprint.

83. A. It is a successor task. The purchasing of the build site must occur before the construction activity begins.

84. B. Plotted points on a scatter chart indicate the type of relationship, or the absence of a relationship, between two variables, one dependent and one independent.

85. C. The access requirements for the project are a guiding reference to what information can be shared and with whom. The complexity in this scenario is that the manager is the functional boss for the team member, and the familiarity may make it tempting to share details about the project.

86. B. Managing project records is an important responsibility of a project manager, who must make sure every document or file has the proper design, format, communication, security is archived and stored correctly.

87. D. A fishbone diagram is used to help determine the different factors that could cause a problem and to do analysis to determine the root cause.

88. A. Make sure each issue has an identifier, an owner, and a due date to resolve to help keep the project on track.

89. C. Strict adherence to the change control process is indicative of a more traditional waterfall approach where the scope is controlled with a more rigid change process.

90. A. A rollback plan is established ahead of time so that everyone knows what immediate action is needed to restore service within the maintenance window. In the middle of a process and or a crisis, it can be difficult to think on the fly the steps that might be needed to restore functionality.

91. B. When project teams are separated by large distances, mountains, canyons, rivers, oceans, or other factors, geographic factors can influence how and when communication will occur.

92. A. Project managers can spend up to 90 percent of their time communicating with the stakeholders and the project team on status updates, getting information, and giving assignments.

93. B. Performing quality assurance would take place during the Execution phase, and managing stakeholder expectations should occur throughout the project.

94. D. This is a project burndown chart.

95. C. The project charter contains the initial expectations of scope, what success looks like on the project, and what objectives the project is trying to meet.

Index

successor tasks, 14, 27, 177, 200, 209, 214, 266, 273
sustainability, 142, 143, 146, 147, 148, 149, 196, 253, 254, 255, 256, 272
SWOT, 128, 161, 166, 248, 261, 262

T

tailored method based on content of message, 31, 215
target audience, 18, 210
task board, 97, 98, 102, 116, 237, 238, 239, 244
task completion, 76, 231
tasks. *See also* predecessor tasks; successor tasks
    total time for completion of, 29, 214
    types of, 14, 209
team building, 30, 161, 215, 261
team selection, 53, 223
team touchpoints, 64, 227
teleconferencing, 195, 272. *See also* voice conferencing
telephone calls. *See* phone calls
telephone conferencing, 98
terms of reference (TOR), 44, 194, 220, 271
testers, 184, 268
testing
    automated testing, 140, 253
    manual testing, 141, 151, 197, 253, 257, 272
text messaging, 6, 89, 98, 106, 206, 234, 238, 241
three-point estimates, 57, 93, 224
tiered architecture, 141, 152, 253, 257
time and materials, 117, 118, 119, 122, 244, 245, 246
time management techniques, 21, 211
time zones, 24, 213
timeboxing, 11, 21, 182, 196, 208, 211, 268, 272
timeline change, 164, 262
top-down estimating, 94, 236
TOR (terms of reference), 44, 194, 220, 271
trademark, 138, 252
transfer, 132, 249
transfer, negative risk strategy, 67, 228
transition/release plan, 49, 54, 55, 56, 221, 223, 224
triple bottom line, 147, 256
triple constraints, 41, 48, 120, 193, 218, 219, 221, 245, 271

U

user story, 11, 51, 185, 207, 208, 222, 269

V

validation checks, 140, 253
vendor knowledge base, 109, 241
version control, 59, 60, 95, 99, 115, 225, 237, 238, 244
videoconferencing, 98, 238
virtual meeting, 9, 119, 181, 207, 240, 267
voice conferencing, 98, 105, 108, 195, 238, 240, 241, 272

W

waterfall methodology, 24, 27, 38, 213, 214, 217
WBS (work breakdown structure). *See* work breakdown structure (WBS)
weak-matrix organizational structure, 14, 208
weather conditions, 27, 214
whiteboard, 88, 89, 234
Wiki knowledge base, 86, 101, 195, 234, 239, 272
Wiki pages, 103, 240
WIP (work in progress), 38, 217
word processing, 97, 238
work breakdown structure (WBS), 15, 43, 48, 50, 52, 54, 72, 75, 144, 178, 179, 187, 193, 209, 219, 221, 222, 223, 229, 231, 254, 266, 267, 269, 270, 271, 1902
work effort estimate, 29, 214
work in progress (WIP), 38, 217
work package, 49, 221
workflow and e-signature platforms, 97, 238
workshop meeting, 13, 208
written, detail communication, 25, 213

X

XaaS (anything as a service), 137, 157, 251, 260

Online Test Bank

To help you study for your CompTIA Project+ certification exam, register to gain one year of FREE access after activation to the online interactive test bank—included with your purchase of this book! All of the practice questions in this book are included in the online test bank so you can study in a timed and graded setting.

Register and Access the Online Test Bank

To register your book and get access to the online test bank, follow these steps:

1. Go to www.wiley.com/go/sybextestprep. You'll see the **"How to Register Your Book for Online Access"** instructions.
2. Click "here to register" and then select your book from the list.
3. Complete the required registration information, including answering the security verification to prove book ownership. You will be emailed a pin code.
4. Follow the directions in the email or go to www.wiley.com/go/sybextestprep.
5. Find your book on that page and click the "Register or Login" link with it. Then enter the pin code you received and click the "Activate PIN" button.
6. On the Create an Account or Login page, enter your username and password, and click Login or, if you don't have an account already, create a new account.
7. At this point, you should be in the test bank site with your new test bank listed at the top of the page. If you do not see it there, please refresh the page or log out and log back in.

SYBEX
A Wiley Brand

Printed and bound by CPI Group (UK) Ltd, Croydon, CR0 4YY

07/07/2023

03233620-0001